THE EXTREME LIFE OF THE SEA

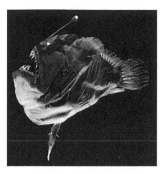

THe EXTREME LIFE OF THE Sea

STEPHEN R. PALUMBI and ANTHONY R. PALUMBI

PRINCETON UNIVERSITY PRESS
Princeton and Oxford

Copyright © 2014 by Princeton University Press

Published by Princeton University Press, 41 William Street, Princeton, New Jersey 08540
In the United Kingdom:
Princeton University Press, 6 Oxford Street, Woodstock, Oxfordshire OX20 1TW

press.princeton.edu

Library of Congress Cataloging-in-Publication Data

Palumbi, Stephen R.
    The extreme life of the sea / Stephen R. Palumbi and Anthony R. Palumbi.
        pages   cm
    Includes bibliographical references and index.
    ISBN 978-0-691-14956-1 (hardcover : alk. paper)
    1. Marine animals.   2. Marine biology.   I. Palumbi, Anthony R., 1984–   II. Title.
QL121.P35 2014
591.77—dc23   2013038174

British Library Cataloging-in-Publication Data is available

This book has been composed in Chapparal Pro with Cyclone and Myriad Pro display by Princeton
Editorial Associates Inc., Scottsdale, Arizona.

Printed on acid-free paper. ∞

Printed in the United States of America

10 9 8 7 6 5 4 3 2 1

Contents

Preface and Acknowledgments   Guiltless Wonder

This book emerged, as many things do, from a gap in the teeth of the established order. The sea is our world's most fertile stage, populated by wonderfully colorful characters acting out their lives in a daily drama. But too often writing about the sea, or any natural habitat, follows this script: detail the native diversity, catalogue species to awe and amaze, and then toll the bell of doom as you explain the man-made calamities unfolding in these habitats. This is the gap in the established order. A gap in character development. And a gaping need for a simple sense of guiltless wonder about how wonderful the ocean's life actually is.

Humans have done sufficient damage to every habitat that the bells can always be heard, but we seek to place emphasis elsewhere. How can an audience focus on a drama's denouement until they're invested in the players? Where they live? Who they live with? The conflict and beauty of their lives? Hence our focus on the characters in the oceans, the lives they lead, and the tactics they use to thrive. We have tried, in our chapters, to bring these characters to life by combining a novel's narrative flair with the scientific accuracy that these subjects demand. And we chose the sea's most extreme life to show what life is fully capable of.

We apologize for inaccuracies that may still be present: despite a world-wide network of friends and colleagues, a survey comprising more than 200 topics will never be perfectly accurate while research is ongoing and new results are arriving. Throughout, we used the scientific literature as the foundation of fact on which our narrative is built—but good storytellers also try to show their subjects in living color, in dynamic movement, and in life and death. And for these elements we sometimes constructed scenes that are fully consistent with the data but may not yet have been witnessed.

We have provided many sources for the material used, but we ran into an interesting problem during our research. Steve, as a Stanford University faculty member with full access to its libraries, enjoys unparalleled access to all the world's published scientific papers. Tony does not. We tried originally to write this book with no more access to the literature than any other Internet user has, to see whether it could be done. But some of those sources are incomplete or inaccurate, and so we switched to relying on the published literature extensively. The difficulty of public access to published, accurate science was frustrating, but in the end we try to serve as that source for our readers and provide a tantalizing citation trail for the reader to follow if she chooses. We have presented online sources when they appear to be accurate, older works past copyright when they are still relevant, and open-source scientific papers where available.

The interface of novelist and scientist, in the end, proved much more interesting to both of us than any father-son dynamic. We hope such pairings may, in the future, render the world of environmental literature more attractive to readers and writers alike. Moreover, we hope you enjoy reading and thinking about the final product as much as we have enjoyed creating it.

The authors thank—for their support, advice, and guidance throughout this project—the wonderful family they are fortunate to share, particularly Mary Roberts and Lauren Palumbi. Lauren Palumbi did the drawings at the beginning of each chapter. We also thank Princeton University Press for their support, Peter Strupp of Princeton Editorial Associates for his relentlessly detailed management of the book's production, and Alison Kalett at the Press for seeing this project's potential when even we weren't entirely sure about it.

We have been fortunate enough to have been helped along by a large number of colleagues, students, friends, and indulgent family members. These include Farook Azam, Scott Baker, Mark Bertness, Cheryl Butner, Greg Caillet, Penny Chisholm, Chris Chyba, Ann Cohen, Dan Costa, Larry Crowder, Mark Denny, Emmet Duffy, Rob Dunbar, Sylvia Earle, David Epel, Jim Estes, Jed Fuhrman, Bill Gilly, Steve Haddock, Roger Hanlon, Megan Jensen, Les Kaufman, Lisa Kerr, Burney Le Boeuf, Sarah Lewis, Jane Lubchenco, John McCosker, Robert Paine, Jon Payne, John Pearse, Dan Rittschof, Clay Roberts, Maggie Roberts, MaryAnne Roberts, Paul Roberts, Sherry Roberts, Carl Safina, Dave Siemens, George Somero, Danna Staaf, Jon Stillman, Dan Tchernov, Stuart Thompson, Cindy Van Dover, Charlotte Vick, Amanda Vincent, Bob Warner, and Craig Young. Each has contributed help, wisdom, warnings, and encouragement.

THE EXTREME LIFE OF THE SEA

PROLOGUE THE EPIC OCEAN

If you stand on a beach and stare out toward the horizon, perhaps squinting at the sunset or the vaporous plume of a distant whale, you can see about 3 miles out. If the weather is clear, you might be looking at 10–20 square miles of ocean surface—a fairly large habitat by most wildlife standards. But the global ocean is actually 10 million times the size of your view out to the horizon, and on average there are more than 2 miles of water under every square foot of surface. The most extreme thing about the ocean is its sheer, inconceivable size.

In that enormous volume—the biggest habitat on Earth—lives a kaleidoscope of animals, plants, microbes, and viruses. Indeed, the ocean nurtures the most fascinating and unique creatures in the natural world. They occupy many different habitats and deploy diverse survival strategies. None lead particularly easy lives. The ocean might seem bucolic from a beach house's front deck, but it's usually too hot or too cold, lousy with microbes, or piled with tier upon tier of predators.

But extreme life thrives in the oceans—whether through speed or guile or infrared vision, by dint of marvelous specialized adaptations. Walt Whitman wrote, "I am large, I contain multitudes" in his "Song of Myself," but the famous line perfectly describes the sea.[1] Dark, deep, and filled with alien creatures, the ocean chills our bones as it stifles our breathing. Its psychological gravity pulls human imagination back to the oldest stories of man struggling against the squalling sea. Our goal in this book is to illuminate the species that have risen to the challenge of the oceans, in the most extreme environments and in the most familiar: the ones using the wildest survival tactics in the sea. We bring you the fastest, the deepest, the coldest, and the hottest, drawing in some of the smallest details of their lives but also painting the backdrop of their role in the oceans.

What lies beneath the sea is more intricate, compelling, and fascinating than the storm-whipped sails of literature or the sensational fearmongering of *Shark Week* television. Most sharks, after all, are not actually that extreme—excepting the few big ferocious ones. Look closely at any plot of water on Earth and a fascinating and awe-inspiring dance unfolds among the ocean's wild denizens. Flying fish flash across the waves with lightning-fast mahi mahi in pursuit. Tropical reefs chatter with the sounds of distant firecrackers as tiny pistol-clawed shrimp fire off powerful sonic weapons. Black dragonfish use infrared vision to ambush hapless passers-by in the depths.[2] Life is a carousel of struggle and success, of beauty and beautiful ugliness.

Over the past few decades ocean science has drawn more eyes toward ocean life than ever before. It's brought more answers to the surface, using an arsenal of scientific approaches and technological instruments to solve mysteries. In 1930, as famed scientist and explorer William Beebe climbed into his bathysphere and dropped into the warm Bermuda seas, he had only an electric searchlight to see in the darkness and a telephone line to describe to the surface what he saw. Today there are submarines, DNA sequencers, robot chemistry labs that skim the waves, and respiration chambers tiny enough to measure the breath of a barnacle. Since Beebe's time we've accumulated more than 80 more years of basic scientific knowledge, without which hardly any serious biological mystery can be solved.

When splashing into the sea with a SCUBA tank and mask, it is hard to predict which of these tools might help you understand something as simple as a bleached coral head. But two things you are certain to need: a delighted sense of wonder at every mystery, and a spark of joy at each discovery, in every creature in creation. Our aim is to give you both.

The world's biggest predator meets its most fearsome prey

It's dark and cold and very deep. A sperm whale (*Physeter macrocephalus*) cruises through the ink, descending toward the floor of the world. He's hunting: powerful muscles and hot blood collaborating to run down rare prey in the cold, oxygen-poor depths. Down and up, dive and ascent, each cycle punctuated with foul-smelling blowhole gasps at the surface. A long life and great bulk lend the bull patience, and he passes by trivial morsels in search of more substantial fare. His broad tail and heavy muscles produce a steady cruising speed. Tiny eyes little bigger than a cow's peer through deepening blues, oriented to look down and not ahead. In the dark, that patience bears fruit: a mile down, the world's biggest predator meets its most fearsome prey.[3]

The silver behemoths known as giant squid measure between 20 and 55 feet in length.[4] Eight short arms are joined by two long, slender tentacles with paddles on the ends: like whips, they're used for hauling prey toward a viciously sharp beak. A typical fish-market squid carries nothing but gentle suckers on its arms and tentacles, but the deep's titans are far better armed. Some have swiveling hooks on the tips of their tentacles; others have serrated suckers like circular saws to rend flesh to ribbons.[5] Prey, so preciously rare in the deep sea, can't be allowed the slightest chance to escape.[6]

Our bull whale follows the same philosophy. Picture the scene: 40 tons of flesh and hot blood colliding with a 30-foot mother squid at 10 feet per second.[7] Though she weighs only 1,000 pounds, much of that mass is pure muscle. The bull whale uses the prow of his skull as a battering ram, perhaps broadcasting a sonic boom forward from the powerful echolocation machinery in his huge head. The squid slows and deploys her arms, spreading them wide and rotating like a parasol in the dark. When they collide, the squid's boneless body absorbs the impact. She rolls with the blow, wrapping her arms around the attacker's head and jaws. Hooks tear long gaping wounds in his skin, layering fresh damage on top of chalky white scars. He's no stranger to this kind of fight.

He feels her arms between his jaws and chomps down, severing two completely. Blue and red blood, from squid and whale, mingles in black water.[8] One swipe of a club-like tentacle knocks out a tooth but does nothing to slow the bull's chewing. Every stroke of his jaws is another awful wound, and for all her fighting spirit, she just can't win. Her razor-edged suckers rip painful chunks from the bull's flesh, but the damage is only skin deep. The squid tries to break away, but half her arms are gone or hanging by ribbons; her siphon works as hard as it can to jet her to safety. But it's not enough. He's too strong, too fast, the fire in his blood fanned by rich oxygen inhaled from the surface.[9] One more strike ends her life in a gush of dark fluids and pink detritus. The whale drifts away, angling his fins like bowplanes toward the surface. What's left of his prey dangles from his mouth as he seeks his next breath of air.

These epic tales are written in scars, published on the skin of victorious whales. The squid sign their names in those scars: colossal squid (*Mesoncho-*

teuthis hamiltoni) leave long parallel gashes, but the saw-suckers of the giant squid *Architeuthis* etch eerily perfect circles.[10] A sperm whale hunting a giant squid has never been directly observed. Instead we read the scars and count the squid beaks in a whale's stomach and know that in the deep basement of the sea, there is a battle going on.

This is not an exotic fantasy. It is a picture that has been built up over a century of careful cataloging and serendipitous encounters. It is matched by an ocean of other battles, from the serial kleptomania of hermit crabs to the war over gonads in sea squirts. It pitches ancient adaptation against today's fight for a meal. It's the sort of thing that happens every day in the extreme life of the sea.

CHAPTER 1   THE EARLIEST

*The history of life is full of fast starts
and odd experiments.*

A poisonous beginning

This planet did not start out as a cradle of life—in its earliest years, it was a hellscape. We wouldn't recognize our world; a time-traveling visitor would need a space suit to survive for even a second. The atmosphere was a thin gruel of carbon dioxide (CO_2) and nitrogen, entirely devoid of oxygen. The ground was streaked with lava, the sky sundered by volcanic lightning. Noxious chemicals bubbled to the surface and into the atmosphere: ammonia, sulfates, formaldehyde.[1] The oceans grew, condensing out of the planet's crust or falling from the sky as rain, but also being delivered piecemeal by incoming asteroids that bore frozen water.[2] And complex chemicals were suspended dilutely in that ice from deep space, seeding the young planet with the molecular materials of life.[3] It was in that alien, chemical stew that the very first elements of life appeared: nucleic acids and proteins. Just a few hundred million years after the crust cooled from magma, life had a lease on Earth.

That life would thrive, but not without crises, and not without experiments and failures and eventual successes. And in those early eons, the oceans swaddled the life of Earth, nurturing it and testing it and setting the conditions for life to persist. Eventually, the living tenants of the oceans grew abundant enough to change its chemistry, altering the very atmosphere of Earth, and building in the sea a complex web of species that exploded in diversity. Life took these skills onto land and transformed that realm as well. But while the cousins of mudskippers colonized the shores and eventually led to human beings, life continued to evolve into nearly every corner of the sea, finding food sources, then becoming food sources, and evolving the abilities to thrive in every kind of environment.

The very first

"Omne vivum ex viva," Louis Pasteur once glibly proclaimed: life springs always from life.[4] Intuitively it seems that the first life would be an exception, but it all depends on your definition of life. The very first self-replicating organic forms weren't organisms per se; they were simply large molecules—molecular machines—and they probably began in the sea.[5]

The process was rapid. The first evidence of life—a signature shift in the isotopes of carbon found in rocks—appears 3.85 billion years ago in the Isua Supracrustal Belt of western Greenland,[6] just 550 million years after the planet's crust finally cooled from magma. Not only did life accrete quickly; it was also sufficiently robust to endure some serious punishment.

The early solar system was a new construction site, littered with asteroids left over from planet building. Careful cataloging of the Moon's craters paints a picture of the rain of asteroids and comets onto our young planet. In those early eons, a series of cataclysmic asteroid strikes had enough power to vaporize the young planet's oceans and sterilize its surface.[7] New early-Earth models suggest that life might have survived these catastrophes, but only if it were already widespread when they occurred. A microbial community that spanned the globe could perhaps hide in deep ocean crevices, buffered from the devastation of planet-killing asteroid strikes and feeding off chemicals bleeding from the molten mantle. Once cellular life had sunk tendrils into ocean habitats as diverse as shallow pools and the deep sea, and once the early solar system was cleared of some of its original debris, life on Earth gained a permanent foothold.

The Great Oxygenation Event

Although life on Earth emerged fairly quickly, it took a long time to evolve past the basics. Recognizable living cells were present on Earth and common enough to form microscopic fossils 3.4 billion years ago.[8] Rocks 3.4 billion years old in South Africa have a series of laminations and filaments suggestive of microbial mats formed in a shallow sea.[9] Yet the world was still devoid of anything but microbes; for two billion years the only living things on our planet were single celled. Their sputtering metabolisms weren't powerful enough to sustain anything grander. Life needed a new kind of metabolic engine to compete at the next level, and it was only invented in response to the planet's first toxic waste crisis. That nasty poison was oxygen, loosed into the atmosphere by the worst of all primordial polluters: photosynthesizing microbes.[10]

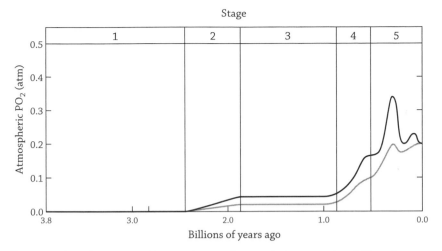

The curves show the upper and lower estimates of oxygen in the atmosphere billions of years ago. Our modern atmosphere has an oxygen concentration (PO_2) of 0.21 atmosphere (atm). Redrawn from Holland, H. D. 2006. "The oxygenation of the atmosphere and oceans." *Philosophical Transactions of the Royal Society B* 361:903–915.

Photosynthesis uses sunlight to form sugars from CO_2. The first forms of photosynthesis arose as early as 3.8 billion years ago and were thought to have been profoundly different from those common today. Most importantly, they did not yet produce oxygen.[11] Oxygen is a home wrecker. The oxygen atom itself binds easily to other atoms, disrupting their chemical bonds. Oxygen atoms insinuate themselves slyly into nearly everything they encounter, breaking bonds faster than after a celebrity marriage. The word "oxygen" is itself derived from *oxys:* Greek for "acid." And because of its disruptive chemical properties, oxygen destroys delicate RNA and DNA molecules, and even disrupts the more stable proteins of cellular life.

Single-celled microbes called cyanobacteria first started eating sunlight and producing oxygen more than two and a half billion years ago,[12] dumping foul *oxys* into an early-Earth atmosphere that was likely nitrogen heavy.[13] Various chemicals in the atmosphere and the soil absorbed that oxygen, "reducing" it and thereby protecting life's fragile foothold. The balance held for a while, but as the cyanobacteria multiplied and oxygen production soared, something had to give. About a billion years ago, oxygen started to accumulate like junk in the garage.[14]

The Great Oxygenation Event was a catastrophe for life on early Earth. Only a few organisms were prepared to make use of the now-ubiquitous poison. But, as *Jurassic Park's* Ian Malcolm declared, life finds a way. It found a way to feed on oxygen, using the crackling chemical energy of its bonds to power a new and mighty metabolic engine. If we think of metabolism without oxygen as a puttering outboard motor, then metabolism burning oxygen is a roaring Ferrari sports car.

Most of what we consider as "advanced" cell features, embodied in a line of cells called the eukaryotes, followed in the wake of this transition to an oxygen metabolism. These features importantly include subcellular organelles called mitochondria, which capture oxygen and burn it chemically to release its energy for the cells' benefit. Mitochondria were once free-living bacteria with oxygen metabolisms:[15] they were co-opted by the cells of our earliest ancestors, and they gifted these cells with the ability to burn oxygen too. Our existence as a species—indeed, the entire organization of life on Earth as we know it—is the unintended consequence of the use of oxygen after this toxic waste dumping.[16]

Archaea

Long before the Great Oxygenation Event, the family tree of life on Earth had its trunk split in two. Of course, both bifurcations consisted of microbes—there was nothing else alive at the time. The first branch was composed of the cyanobacteria and other "normal" bacterial microbes. The second branch emerged around the same time, made up of microbes evolving to endure in stressful environments[17] or living on a chemical diet without sunlight. They are Archaea—the extremophiles—the toughest creatures ever to live.[18]

They're nothing much to look at: tiny oblong masses beneath an electron microscope. For a long time, we thought they were just bacteria. With the advent of gene sequencing, biologists noticed a huge genetic gulf between these extremophiles—found in salt lakes and deep-sea sulfur vents—compared to typical bacteria. In response, taxonomists created a name for this entirely new domain of life: Archaea.[19] More and more of these creatures have been discovered at the margins of the world, living in places where little else could survive: the hot springs of Yellowstone National Park,[20] the hydrothermal vents on the floor of the ocean, the oxygen-poor deep sea. As creatures of early Earth, Archaea tend to get crowded out by latter-day microbes. So they remain in extreme environments, in the closest analogues to the planet they lost.

Archaea can grow at temperatures exceeding 230° F (110° C), well above the boiling point of water. The Archaean *Pyrolobus fumarii* coats hydrothermal vents, 6,000–8,000 feet below the ocean surface, where sulfur and other toxic chemicals spew from Earth's crust at temperatures of hundreds of degrees. These creatures hold the world record for growth at a high boil. They can survive an hour at autoclave temperatures of 250° F (121° C) and find temperatures of 203° F (95° C) too cold to reproduce well.[21] No multicellular animal or plant can grow at such temperatures (see Chapter 8), and so the hottest places on our planet are solely tenanted by microbes. But microbes used to rule not just in high heat, but everywhere.

The Cambrian Explosion

For a long moment in evolutionary history, across our entire planet, tiny microbes qualified as the most complex organisms on the planet. Eventually, some were able to form larger structures: thin layers of bacterial cells and secreted limestone piling one atop the other into mounds called stromatolites that persist three billion years later. These were still microbial constructions; no organism bigger than a single cell existed on Earth for eons.

Precisely when or how the jump was made from microbes to animals is not recorded. Fossil records are notoriously patchy, like the picture on an old television set. Static swells the further back you go. Large jelly-like organisms of many species all lumped together under the generalized name Ediacara appear in 575- to 542-million-year-old mud deposits.[22] Other early cell clusters look like embryos of large creatures, though they might just be groups of single-celled protists.[23] Tiny swimming discs like the bells of jellyfish wafted through the sea. On the floor were soft organisms that looked like disks, bags, toroids, or quilts.[24] Whether these lines diversified into the life on modern Earth is ultimately unknown. They could be failures—snipped-off stubs on the evolutionary tree—or they could be the ancestors of all current animal life.

Our understanding of these early experiments in multicellular life completely changed in 1909 when paleontologist Charles Walcott walked into a quarry in British Columbia, Canada. He stood agog: before him spread an ancient deposit of ossified mud, more than 500 million years old and about the size of a city block. Dubbed the Burgess Shale, it remains to this day the planet's best-preserved record of ancient marine organisms. This hunk of old ocean floor might be the most important discovery in modern paleontology.[25] It documented, for the very first time, a worldwide biological revolution.

This enigmatic Ediacaran fossil represents one of the first multicellular species. But whether it is an animal, a fungus, a lichen, or something entirely unlike life today remains a mystery. Photograph by Meghunter99.

Cataloging a fossil dig like the Burgess takes a long time and an enormous amount of work. As Walcott, his family, and an army of paleontologists mined away and meticulously recorded their findings, based on more than 65,000 specimens, amazement with the newly discovered species grew. The creatures of the Burgess Shale were hard to fit into the normal taxonomy of living invertebrates. Their bodies were like unique jalopies assembled from random spare parts. Odd feeding trunks, long spiny legs, bony lobed fins, the wrong number of compound eyes—all these and more were trapped in the mud, slapped haphazardly onto animals that looked more like sci-fi cartoons than actual living creatures.

Wiwaxia was a small, scaled slug-like creature studded with petal-like flaps. *Marrella* resembled a brine shrimp wearing a motorcycle helmet, trailing long graceful tentacles from its mouths past its tail. *Odaraia* looked a bit like a fish in a hot dog bun, its torpedo-shaped body bulwarked on either side with large translucent carapace shells. Two compound eyes made of many facets, tiny feeding appendages near the mouth, and a bizarre three-finned tail rounded out *Odaraia's* alien appearance.

The modern invertebrate world has long claimed its share of anatomical oddities, which biologists and taxonomists have labored hard to catalog. The Burgess Shale tossed a whole new menagerie on the table and threw taxonomy into a tailspin. All this new variety showed that the Cambrian was a unique period in natural history, when countless different forms leapt into existence in a geologic eyeblink: a worldwide phenomenon dubbed the Cambrian Explosion.[26]

The empty barrel

The Burgess Shale is a snapshot taken about 505 million years ago. It captures the event in full bloom, the seas writhing with different body forms that all appeared in an extremely short period. Suddenly, without any obvious precipitating event, the sea had birthed advanced life at an evolutionary speed previously thought impossible. How could so many disparate creatures appear with no prior hint from the fossil record?

Darwin had died almost 30 years before Walcott walked into the Burgess Shale, but his evolutionary theories combined with twentieth-century science point to possible origins of the Cambrian Explosion. One satisfying theory, as popularized by the famous paleontologist Stephen Jay Gould, is called the empty barrel. The theory starts with the very first life: it was microbial, because it had to be. Low levels of atmospheric oxygen limited metabolism and created an environment hostile to large multicellular life forms. When at last there was enough oxygen to fuel powerful metabolisms, more complex creatures found themselves in much more benign environments, full of bacterial food and the metabolic energy needed to thrive.

Propelled by vast supplies of bacteria to feed on, and initially freed from competition's cruel shackles in an ocean empty of animal life, evolution went berserk. Natural selection is supposed to weed out inefficiencies, one of Darwin's ideas, but in the early Cambrian oceans any misbegotten jumble of genes and components capable of feeding itself would find success as a multicellular animal. There was almost nothing so absurd that it couldn't survive *somewhere*. The oceans were an empty barrel, and life filled it with extreme speed and astonishing diversity.

Despite horrific extinction events in the intervening half-billion years, the marine world has never again been so empty. The Explosion's products were here to stay, delicately twisting one another's genomes through competition and predation. Ecology's niches were always at least partially occupied; populations rose and fell but never completely vanished. As Andrew Knoll notes in *Life on a Young Planet*, "no one who has trekked through thick successions of Proterozoic shale or limestone can doubt that Cambrian events transformed the Earth. Cambrian . . . evolution may have taken 50 million years, but [that era] reshaped more than *3 billion* years of biological history."[27]

Inventing predation

Besides the empty barrel, there is another plausible theory to help account for the Burgess Shale's sudden diversity. It springs from new data from a very different kind of biological research: genetics. Advanced gene sequencing opened up new avenues of research and helped provide a different way of estimating when major taxonomic groups evolved.

When two groups diverge, their genes begin to accumulate different mutations. If we know the rate at which these mutations build up, and we know how many mutations separate the two groups, we can estimate how long ago they diverged. This mechanism, called the Molecular Clock, is a way to measure time through DNA, counting tiny mutations like ticks of the second hand and projecting eons across the evolutionary gap from starfish to lobsters.[28] It tells us that the DNA of modern invertebrates is too different from one phylum to the next to have started accumulating mutations from the time of the Burgess Shale. Instead, the different types of invertebrate life must have diverged millions of years before—perhaps hundreds of millions.[29]

But if life has such a deep history, if arthropods and mollusks existed before the Cambrian, then why didn't these different forms appear in the fossil record? Several possibilities might occur to you: (1) they were there all along, and produced fossils, but we have not found them yet; or (2) they were there but *didn't* produce fossils, because they didn't have body structures for fossilization.

Skeletons produce the best fossils. Soft tissues rot quickly, but hard parts—crab carapaces, snail shells, and fish bones—endure a long time. Suppose the ocean before the Cambrian was full of half-inch octopuses—would we ever know? It turns out that octopuses have hard beaks made of a protein similar to fingernails that can make good fossils, so we *would* know. But what about a half-inch snail without a shell, or worms burrowing in mud, or tiny starfish with soft fleshy arms? None of these would leave more than a trace of movement left behind in the sand.

In fact such trace fossils have been abundantly found, starting about 600 million years ago.[30] Complex three-dimensional burrows of worms and scratch marks from the world's first scrabbling appendages are etched into medallions of mud. And they predate the appearance of the complex skeletons of the Cambrian Explosion. So the explanation partly may be that many kinds of creatures existed before the Cambrian Explosion, but without fancy hard skeletons. Then what happened? Perhaps it was the planet's first arms race.

As animals evolved into larger, mobile forms, they moved from grazing on helpless microbes to feeding on one another.[31] Predators needed sharp claws and strong limbs to shred their prey. That prey, unwilling to be dinner, evolved defenses: shells, carapaces, spines, and toxins. In turn, predators countered with stronger jaws and teeth. And in response, prey evolved more defenses. Bit by bit and stage by stage, the animals may have evolved the weapons and the armor and the defenses that we see so abundantly in the fine fossils of the Burgess Shale. An explosive arms race that did not destroy life but diversified it.

A portrait of winners and losers: *Pikaia* and *Opabinia*

The Burgess Shale is an antique fair of life, packing thousands of species into a modest mud parking lot. Over years of analysis, paleontologists have isolated a handful of creatures they thought most fundamental to life's later story. Some were winners, and others were magnificent losers. But they all thrived for a time. Their disparate features illustrate just how unpredictable evolution can be when left to run wild.

In 1911, Walcott named a creature he thought of as a segmented worm, calling it *Pikaia gracilens* to convey the grace he imagined in its movement. About 2 inches long and delicately segmented, it was lumped together with about thirty other worm fossils that were later sent to a young Englishman named Simon Conway Morris. But Conway Morris, an early practitioner of three-dimensional fossil excavation techniques, quickly realized this was no worm. A hard central structure ran from one end to the other, girded on either side by zig-zagging bands of muscular tissue. Conway Morris was looking at a primitive backbone; a feature that would eventually grow into the crucial pillar of our nervous systems. The banded patterns in *Pikaia* were muscles

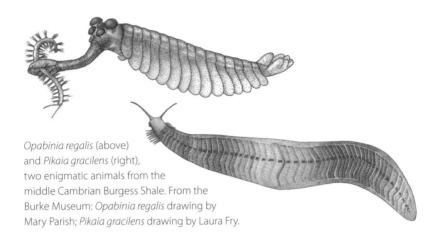

Opabinia regalis (above)
and Pikaia gracilens (right),
two enigmatic animals from the
middle Cambrian Burgess Shale. From the
Burke Museum: Opabinia regalis drawing by
Mary Parish; Pikaia gracilens drawing by Laura Fry.

bearing more resemblance to the repeated backbone structures of vertebrates than to anything ever seen in a simple worm.[32] Precisely what *Pikaia* did in those seas of 500 million years ago remains a mystery, but it was fairly rare and disappeared not long afterward. Gould doubted that *Pikaia* was a universal vertebrate ancestor, but it seems to be our closest Cambrian relative. For Gould, *Pikaia* and its relatives were the thin line of life that perpetuated our kind from the heart of the Cambrian Explosion to the rest of evolution's theater. This thin proto-spine in the Middle Cambrian suggests that the vertebrate body plan—so dominant today—was just one spinning spot of color in life's early kaleidoscope.[33]

Compared to *Pikaia*, *Opabinia regalis* is an even bigger Burgess oddity. Though only a few inches long, the unveiling in 1972 of a new drawing of it to the Oxford Paleontological Association was met with incredulous laughter.[34] *Opabinia* looked so bizarre that even the assembly of sober scientists could react no other way. The animal is oblong, a slightly engorged cigar trailing frond-like gills from the top of its carapace. A frondy tail propelled it slowly along the bottom, while five eyes (rare among arthropods, which tend to pair off their body parts) sat on stubby stalks. The gut ran tail-to-head in familiar fashion but took a bizarre U-turn at the head to form a backward-facing mouth. And protruding like a segmented vacuum hose from the head was a thin prehensile trunk. The nozzle-like device was tipped with spiny pincers, but it was also the perfect length to reach back to the mouth. *Opabinia* likely trolled the muddy bottoms of the Cambrian sea, rooting in the muck for tiny creatures.[35]

As in the case of *Pikaia,* early researchers tried to shoehorn *Opabinia* into a clean taxonomic bracket. Walcott considered it the perfect early arthropod![36] But later reconstructions by Oxford paleontologist Harry Whittington broke the shoehorn by showing the little creature could not fit into arthropodan preconceptions. *Opabinia* stubbornly resisted classification and essentially has been left out of the hierarchy of life.

Stephen Gould, among others, saw *Opabinia* as a stunning reinvention of the conventional: the key to understanding the entire Cambrian Explosion. Put simply, *Opabinia* resembles nothing in the modern world. Its features are so disparate, so far afield of what was expected in an arthropod, that it forced paleontologists to discard their assumptions about what was necessary for an organism to be successful. The residents of the Burgess Shale, unusual as they seem today, lived and fed and fought and died just as today's creatures do. Which one was part of a line of animals that survived and thrived, and which one was in a group that dies out might have been preordained by their basic biological features. But maybe just as likely is that things didn't have to unfold this way.

Replaying the game tape

One of Gould's largest contributions to popular evolutionary thought was his emphasis on the role of chance. Gould loved to talk about cathedrals and to use sports metaphors, and in *Wonderful Life* he keeps returning to the idea of "game tape": the idea that natural history, were it to be replayed, might unfold very differently. Consider a famous football game: Super Bowl 44, played in February 2010, when the New Orleans Saints defeated the Indianapolis Colts. If you watch a replay of the game (a record massively more complete than any fossil bed), you can state with confidence that the Saints were the better team. This touchdown, that catch—you can point to every event that led to the natural result: 31-17, New Orleans. But of course, anyone could rightly object: "That was one game." If the game were replayed, any number of things might be different.

Let's examine one play from that game. Following The Who's colorful halftime performance, the Saints attempted an onside kick to try to steal a possession. Onside kicks are typically desperation maneuvers, used when a team is losing with little time on the clock. They fail more often than not. This time should have been no different; the ball took an impish hop away from the Saints and directly into the waiting arms of a Colts player. Yet he allowed it to bounce off his facemask, losing control and ultimately ceding possession to the Saints. It was a huge play, swinging momentum and placing the Saints in control for the second half. But if those events were replayed a dozen times, how many times would the Saints have come up with the ball? If the Colts recovered with great field position, wouldn't Peyton Manning have led them to an immediate touchdown? It's impossible to say.

The Play of Life is enormously complicated. Success and failure hinge on a staggering number of different factors, the number and importance of which cannot be known in advance. For that reason, the outcome will always appear "natural" despite being (more or less) capricious. A meteor impact, an algal bloom, or a temporary El Niño climate swing might snuff an otherwise promising species out of existence. *Pikaia* might have perished and taken all of human history with it. Gould looked at the Burgess Shale's incredible fossils and saw them for what they are: one small snapshot of an enormous game already in progress. The winners of the Cambrian Explosion evolved into the species on Earth today, while the losers retreated as stunted relics squashed in the mud of ancient stones. What if the winners and losers were reversed? Gould explains in his own words:

> any replay of the tape would lead evolution down a pathway radically different from the road actually taken. But the consequent differences in outcome do not imply that evolution is senseless, and without meaningful pattern; the divergent route of the replay would be just as interpretable, just as explainable *after* the fact, as the actual road. . . . Each step proceeds for a cause, but no finale can be specified at the start, and none would ever occur a second time in the same way. . . . Alter any event, ever so slightly and without apparent importance at the time, and evolution cascades into a radically different channel.[37]

Diversity and decimation

All the hundreds of thousands of crustacean species on modern Earth can be grouped into four major groups. They are as different as crabs and sea monkeys (*Artemia,* brine shrimp), though they share a set of common traits like paired legs and segmented muscles. In the Burgess Shale—a spot of land so small that it is just one mudslide the size of a city block in one tiny corner of an enormous planet—crustaceans have been classified into *twenty-four different groups.*[38] Multicellular life poured down many channels, but then subsequent evolution took a few of the most successful anatomical designs and used them as the basis for everything that would follow.

The Explosion's creative chaos was pure and open; then the process of extinction broke life's broad river into the tributaries we see today. *Opabinia* was a casualty of this process, perhaps no less deserving of an evolutionary future than *Pikaia*. But the tape played out as it did, and we (the distant beneficiaries of evolution's early winners) get to wonder at the losers in studies of ancient graves.[39]

Gould makes the point: "The sweep of anatomical variety reached a maximum right after the initial diversification of multi-cellular animals. . . . Compared with the Burgess seas, today's oceans contain many more species based upon many fewer anatomical plans."[40]

The winners didn't necessarily boast the best long-term solutions: maybe they were just the best for a short, special spurt. But then the winnowing began: the initial explosion of genetic creativity was bled white by the eons of extinctions and environmental changes that followed. After the Cambrian Explosion populated the world with odd and complex organisms—with wild versions of experimental life—the slow grind of competition and natural selection killed most of them off. And the winners began to diversify into the myriad related life forms we see in the oceans today. So even though a huge number of different body plans exists now, filling all the various tiny niches in the world's countless environments, they pale in comparison to early Cambrian fauna.

The only predictor of life is . . .

Combine the two Gouldian analogies of the empty barrel and the game tape. The empty barrel means that extreme life forms were simply tolerated in a world that had not yet filled with big, complex life. The game tape analogy reminds us that when exuberant life was evolving with little competition in an empty ocean, extreme creatures are as likely to succeed as any other. What appears on life's stage can surprise us, and many things seem impossible until suddenly they stand before our eyes. We may call these forms unlikely, but that is just because we do not see them in the modern world. And it tells us that what we have come to find "likely" are the forms that we see, not the forms of life that necessarily must be.

Every creature we see today has some kind of root in the Cambrian Explosion. Life originated in the oceans, was sheltered by the sea through cataclysms, and only much later was anything sufficiently advanced to shuffle free of its fluid embrace. In the tens of million years of the Cambrian Explosion, nature and evolution amply populated the stage. The Cambrian Explosion produced the oceans first superstars, the trilobites that dominated the seas. It forged the different life forms that would take over in succession: the cephalopod mollusks that would become the ocean's fiercest predators 400 million years ago, and even the first rough drafts of our own vertebrate body plan. The march of complex ocean life had begun.

CHAPTER 2 THE MOST ARCHAIC

Living fossils use old body plans that still work today.

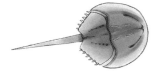

The Volkswagen Type 1, with its walleyed headlights and domed roof, is uniquely enduring among automobiles. An enormous commercial success from its first production in 1938 through the 1960s, the iconic car continues to be driven widely and to inspire enthusiasts around the world. By any objective standard, newer models are better—faster, sleeker, safer, and more efficient or maneuverable. And yet the Volkswagen "Bug" maintains a tenacious grip on existence.

Type 1s may have gone into production more than 70 years ago, but they offer features and characteristics that remain attractive to this day, because they solved common problems. Air cooling reduced the size and weight of the engine while also bringing down the number of components that could break. Combining rear-wheel drive with a rear-mounted engine was excellent for driving in poor weather—the engine's position over the drive wheels weighed them down and offered the vehicle great traction despite its light frame.[1] Finally, the legendary *Sicherheit* of German engineering, combined with a reasonable price, rendered the Type 1 a secure bet for working-class families around the globe. For a while the cars were everywhere. So even though an old Beetle sputtering down the road isn't much to look at, its original features keep it competitive in a more advanced world.

In the same way, it is their features that set apart the planet's "living fossils": organisms that evolved hundreds of millions of years ago and that have succeeded across eons without major changes. Like the Volkswagens of decades past, they tend not to be terribly common now, but they continue to survive in spite of a world that has passed them by. All the ancient organisms surviving to the present day have done so because they possess certain core features that not only ensure their survival but also continue to define them.

Some of these living fossils have always been rare. But some had a rockstar past: they've been among the dominant life forms in the oceans, filling the shallows and the deeps and controlling the ecology of early ocean ecosystems. Yet the evolutionary wheel eventually turned, and newer forms came to dominate our seas. When we look at today's ocean, it seems so normal that it is filled with fast swimming fish, gigantic whales, and leaping dolphins that it might be a shock to discover that prior oceans didn't look that way.

Trilobites rule

As the world heated and cooled, as continents broke apart and then came together like gigantic pools of mercury, before the land was colonized by complex life, the trilobites thrived. They could be found in every ocean, filling various ecological niches: predators, prey, scavengers. They were bottom-crawling, multi-legged creatures with strong carapaces and sharp claws. Among them were the eurypterid sea scorpions measuring up to 8 feet long and bearing barbs at the ends of their tails.[2] These primitive arthropods were the common ancestor of today's spiders, scorpions, ticks, and horseshoe crabs. They shared the trilobites' powerful survival tools and enjoyed similar success.

Trilobites were not Volkswagen Beetles—they were an entire fleet of diverse organisms, a whole production line and not just a single model, that burst upon the ocean scene in the Cambrian, 540 million years ago. By the time of the Burgess Shale (see Chapter 1), they had evolved into a riot of species with specialized spines, legs, and sophisticated eyes.[3] Their exquisite preservation in rocks opens up their world to us in a way that few other fossil groups have ever done. There were "predators, mud-grubbers and filter feeders"[4] that divided up the early food resources of their ancient seas. They rolled up when disturbed, like a panicky pill bug, making spiny-carapaced species that resembled armored pincushions. They left behind their skeletons when they molted, just like crabs do today, and nineteenth-century paleontologists painstakingly put together growth series from tiny juveniles to full adults.[5]

Perhaps the most startling adaptation of these early arthropods were their eyes:[6] large compound structures containing hundreds of small calcite lenses.[7] Calcite is a form of calcium carbonate (the same material as a crab's shell or the sand on a tropical beach) that is usually as transparent as a brick.

But the calcium carbonate in a trilobite's eyes is transparent, "constructed of calcite (crystals) so precisely oriented crystallographically that they behave optically as if they were made of glass."[8] These animals took an opaque skeleton, made it transparent, and watched the world through these unique lenses for hundreds of millions of years. Then they took this crystallographic magic to the grave with them.

A list of the trilobites from the late Cambrian, 500 million years ago or so, is an impressive collection, but the best would come later, in the Ordovician era 490–445 million years ago. In this period, trilobites occupied the most regions of the sea in "sunlit reefs and in gloomy abysses."[9] Later epochs saw trilobites dwindle as one extinction event after another reduced their diversity and numbers. Eventually, trilobites and sea scorpions died out. They disappeared from the oceans about 250 million years ago at the end of the Permian era, when 96% of the ocean's species disappeared in the biggest mass extinction of all time. Five genera of trilobites were going strong until the mass extinction, and paleontologist R. M. Owens mused, "But for the extreme stresses that the entire marine biota suffered towards the end of the Permian, trilobites would probably have survived for much longer."[10]

Trilobites were not a flash-in-the-pan invention. They were a dominant part of the life of the ocean for 200 million years, 100 times longer than our own species has existed. They were the first mega-successes to climb out of the empty barrel of the Cambrian, and they made an empire of the entire ocean. They may have fed the rise of ammonoids, nautiloids, and sharks, and they may have competed with upstart newcomers like fish. But while they were here, they defined the life in the sea. Yet they were living fossils 250 million years ago, not today, because their line completely died out. Other lines were similarly successful but have provided us a few key living species to remember them by.

Chambered nautilus

The distinctive, spiraling rose-colored shell of the chambered nautilus, *Nautilus pompilius,* epitomizes the sea's delicate beauty. It is usually shown cut in half, revealing curved chambers bathed in a pearly glow. Lined with pale lustrous nacre, the shell's interior is a study in mortal humility. Building from a tiny solid core, the animal secretes a curled tube of material with which to shelter its soft body. As it grows, so must the tube—ever wider and longer, looping back around itself again and again.

Chambered nautilus. Photograph by Chris 73 / Wikimedia Commons.

The animal that makes this shell is not so delicate looking: a hulking, tiger-striped creature stuffed sloppily inside a coil of plaster white. A broad, visor-like mantle protrudes from the shell and sits above a writhing clutch of short tentacles. Two alien eyes twitch on the sides of its head, using primitive pin-hole apertures instead of lenses to search for prey and enemies.

Ancient seas held huge numbers of these creatures: nautiloids plus a related group called ammonoids, now extinct. Hundreds of millions of years before the rise of the sharks and fish, these shelled, tentacled creatures were the most advanced predators of their age. They ranged in shell size up to nearly 10 feet across.[11] A shell 10 feet wide probably made ammonoids cumbersome, so they most likely preyed on slow bottom-dwelling creatures such as trilobites. Some had shells twisting around in such strange contortions that it's difficult to imagine how they swam at all. These strange oversized beasts persisted in great variety and huge numbers for hundreds of millions of years.

The heavy shell is the nautilus's defining feature: a legacy of its ancient family line. The chambers are no simple curiosity, no crude solution to a simple problem of unused shell space. As the animal grows, the shell grows as well, and the animal must keep its position at the opening. So it vacates the space where it used to live, moving forward in the shell, and sealing the old apartment off with a thin film of pearly material, creating the "chambers" for which it is named. The chambers themselves are marvels, laid end-to-end to describe a nearly perfect logarithmic spiral.[12]

The chambers are crucial to the animal's survival, necessitated by the shell itself. Thick shells offer security—but at the cost of heavy weight. With such a load, the nautilus should simply sink to the bottom of the ocean. So why doesn't it? The chambers are the answer. They may be vacant, but they're not empty.

The nautilus alters its buoyancy the same way a submarine does: by filling or emptying its ballast tanks. Through each chamber runs a fleshy thread called the siphuncle, connecting the chambers like locks in a canal. This thread alternately oozes fluid into the chambers or absorbs it, altering the animal's buoyancy like a submarine maintaining its ballast tanks.[13] The shifts are gradual, but fast enough to let the animal traverse hundreds of feet up and down the water column as day fades to night. During the day, they bide their time on the darkened slopes of deep coral beds nearly 1,000 feet down. At night, they climb into shallower water to feed. This arrangement has a serious weakness, though. *Nautilus* cannot add gas to pressurize its chambers, and when there is high external pressure—one experiment suggested a threshold around 1,600 feet deep—the shell collapses.[14]

The nautilus navigates the world in the same way as most other cephalopods: by using an elongated tube called a siphon that can eject water from the mantle cavity. By taking in and expelling water, the mantle and siphon acts as a primitive underwater jet engine (like rolling your tongue into a tube to expel air).

For 400 million years nautiloids and their now-extinct ammonoid cousins reigned. The ammonoids died out when the dinosaurs did, about 65 million years ago. The nautiloids also dropped in numbers, finally dwindling to the six species alive today. Like trilobites, their history is one of profligate diversity, ecological diversification, and great fossil longevity. They were sometimes predators and sometimes prey; fossils exist bearing gaping shark bites and other battle scars. Mosasaur bites (large crocodile-like extinct reptiles) have been found.[15]

Hundreds of millions of years ago, a time when nearly all complex animals were bottom-dwellers, an organism of the Nautilus' versatility ruled the seas. The Chambered Nautilus survives to this day because of the same adaptations that made it successful in the first place: the jet-producing siphon and the chambered shell.[16] Hidden away in the depths of isolated reefs, the nautilus is a king "bounded in a nutshell."[17]

Horseshoe crabs

Ocean City, Maryland, is a tectonic boundary between nature and artifice. Built on a spit of a sandbar on the sandy mid-Atlantic coast, the city is in constant conflict with the sea. On the western edge, river waters pool in a brackish estuary with marsh grass growing down to the shore. On the eastern edge, the Atlantic Ocean beats relentlessly on silicate beaches crowded with summer tourists. High-rise hotels and apartment buildings crowd the shoreline, daily disgorging hordes of beachgoers. At the southern end stands a glittering entertainment strip, complete with video arcades and a Ferris wheel embodying a particular, well-aged vision of American glory. Few places are more iconic than an aging Atlantic coast boardwalk: parents explaining to their unimpressed kids *how great this was,* omnipresent stickiness beneath your feet, and storefront tchotchkes appealing to nostalgia and affronting good taste in equal measure.

The boardwalk looks old, though it has been there a scant century. But along the long, sandy beaches is one of the most enduring life forms on the planet. Just beneath the waves, among the sand bars, lives a true "living fossil," known colloquially as the horseshoe crab. After hundreds of millions of years on the planet, these animals remain true to a body form that we can find in the deep rocks of ancient fossil beds, with hardly any substantial change.

The best way to learn about horseshoe crabs is simply to find one on the beach and turn it over. Be warned—though the animals are harmless, most beachgoers are unprepared for the clawed shock beneath. Ten long, spindly, claw-tipped legs reach and curl like dead calcified fingers. They emerge from a glistening insectoid thorax in front of a heavy abdomen and behind a thin head adorned with tough, claw-tipped feelers and topped by a carapace like an alien helmet. Lacking jaws or teeth, horseshoe crabs grind up food with their legs and shuttle it to the tiny sucking orifice. These creatures look like nothing so much as aliens from another planet.[18]

Even in the modern world, they are alone—isolated by eons of evolution. Horseshoe crabs are not crabs in the proper sense. They're more closely related to spiders than crustaceans, properly in the subphylum Chelicerata rather than in Crustacea, and there are only four species left in the world.[19] All have a dome-shaped hard carapace sheltering the entire body. It's less a suit of armor than a rigid tent laid over a separate body. They breathe through leaves of paper-thin respirating tissue, folded over like the pages of a book between the legs and the long nail-like tail. Attached to their back legs, these "book gills" allow horseshoe crabs to breathe without proper gills. These adaptations were common in ages past, but today the horseshoe crab is the only animal on Earth to feature these anachronisms.

The design may be outdated, but there are ways in which horseshow crabs outdo many of its later cousins. For example, the eyes of modern horseshoe crabs are relatively simple: two large primary eyes, and seven others of varying sensitivity and size arrayed across the body. Yet recent research shows the sophistication of vision in these ancient creatures and how their simple brains process complex visual signals. In particular, horseshoe crab eyes are a million times more sensitive to light at night than during the day. Hypersensitive eyes ramp up the sensitivity of the retinas at night and turn them back down during the day, perhaps to pick out mates during night-time low tides.[20]

Despite the primitive features of horseshoe crabs, gills still deliver oxygen to the blood, the ten legs sift sand for food, and males lumber after females to hold them in a fertile embrace. New generations of the same model have been produced down the eons, natural selection is satisfied, and the shape and habits remain unchanged in the four remaining species. Like the Volkswagen Beetle, they stick out like a sore thumb but persist in sturdy practicality.

The Atlantic horseshoe crab is a relatively new species,[21] but horseshoe crabs as a family first appeared on Earth some 450 million years ago.[22] Fossils from 445 million years ago have the characteristic domed carapace of modern horseshoe crabs, along with their thick tails.[23] Long persistence with little physical change is what makes horseshoe crabs the poster children of living fossils; their detailed body plan has persisted longer than just about any other. They had cousins with slightly different body forms over the ages, and some had been stronger and more successful for short spells. But ultimately all the others died out, succumbing to the march of time.

The last trilobite

There are four species of horseshoe crab left, and no trilobites. So even though trilobites ruled the oceans—and horseshoe crabs never did—in terms of survival, horseshoe crabs have won. But there is one type of trilobite left, embodied in the pill-bug-like youngsters of the modern horseshoe crab. Hatched from eggs buried in sand, the "trilobite" larva of the horseshoe crab is the only living vestige of the trilobite dynasty. So horseshoe crabs are actually double living fossils, embodied in their adult forms and in their youngest stages.

Why horseshoe crabs avoided the extinction that consumed their more successful cousins is unknown. But they provide a link back to the ancient seas before fish or whales or modern corals, and they live today on a few protected sandy beaches of a more crowded world of cotton candy and roller coasters.

Coelacanths

What would it be like to be the one who discovers that trilobites are still alive? No one can claim the fame, but exactly this drama played out for another iconic living fossil—a familiar animal from the fossil record, for centuries thought extinct and then amazingly rediscovered! It's the coelacanth, a deep-dwelling rarity whose discovery laid bare the limits of human knowledge.

Coelacanths take their name from the larger order of "lobe-finned" fish to which they belong, an order that appeared a little later than the earliest sharks—about 400 million years ago. There were never many species, but at least one had a large range from North America to Asia.[24] Their thick, bony, finger-like "lobe fins" were relatively primitive swimming instruments that were the evolutionary progenitors to vertebrate limbs. Every bird, reptile, and mammal alive today is descended in some way from this ancient stock of marine ancestors. Paleontologists knew this for a long time, and big fossilized coelacanths were immortalized in more than a few museums. They also knew that the last coelacanths died out in the Cretaceous period, some 65 million years ago.[25] They knew this until 1938, when a young woman bought a fresh-caught coelacanth in a South African fish market.

Dr. Marjorie Courtenay-Latimer recollected the day the coelacanth returned: "22 December 1938 dawned a hot shimmering summers day. . . . [The] phone rang to say the trawler *Nerine* had docked and had a number of specimens for me. . . . So I rang for a taxi and went down to the fishing wharf."[26]

Latimer was serendipitously handed a creature that would one day bear her name: *Latimeria chalumnae*, a large, oily fish trawled up from about 200 feet deep in the Indian Ocean, off the coast of South Africa. For ichthyologists,

this was like finding a live dinosaur in the Amazon. The specimen was immediately taken to Europe, where it was displayed in front of a reported crowd of
20,000 eager science enthusiasts. Latimer and the scientists with whom she
shared her discovery were worldwide celebrities, and the history of marine
evolution had to be rethought. In the case of the coelacanth, countless fishermen must have encountered them, especially along the coast of the Comoro
Islands near Madagascar, where most specimens have subsequently been
found. But between their poor commercial value and relative rarity, European
scientists never made the connection.[27]

The fish themselves seem not much changed from their primordial ancestors. Thick, fleshy fins and a heavy body render the coelacanth slow and cumbersome even in its natural habitat. A tiny brain occupies less than 2% of its
skull cavity. The rest is filled with fat for buoyancy. Its flesh is dense, oily, and
foul with urea—in fact, the species' utter lack of commercial value may be one
reason it still exists. Coelacanths swim ponderously, like underwater zeppe-
lins, and float quietly in wait of the small fish that are their prey. Broad, flat
fins sway back and forth unsteadily as though the fish were about to roll over.
If a shark is the evolutionary equivalent of a serrated butcher's knife, a coelacanth is a wooden club.

Coelacanths remain on the razor's edge of extinction, sustained probably only through isolation on the periphery of life. These fish likely survived
through little more than evolutionary serendipity. They inhabit a narrow
zone in cooler water, between 300 and 600 feet beneath the surface along
steep volcanic coasts, seldom coming close to the surface unless cold upwelled
waters bring them up. They produce just a few, very large eggs—a 40-inch
fish will have eggs the size and hue of an orange. Their ancient body plan still
makes them an effective predator of more modern fish, but they remain present in only a tiny fraction of the world they once inhabited.

Sharks at last

Beautiful, powerful, and thrilling, sharks have ever gripped the human imagination. They are, it has been said, eating machines; perfectly honed instruments of death, the Platonic ideal of the cruel predator. In her book *Demon
Fish*, Juliet Eilperin quotes the Greek poet Oppian, describing sharks "rav[ing]
for food with unceasing frenzy, being always hungered and never abating the
gluttony of their terrible maw."[28] Their bodies are streamlined torpedoes of
hard muscle, equally suited for effortless gliding and the explosive speed of

the kill. Their mouths are not tools, but weapons: rank upon rank, teeth like serrated knives advance to replace those shattered on the bones of last week's lunch. Outlandish sensory organs allow them to detect tiny amounts of blood over vast distances and "hear" the electrical pulses of a victim's heart. Denticles on the skin—specialized scales that resemble nothing so much as tiny sharp teeth—create minuscule low-pressure pockets of water along the body. Like divots on a golf ball, the pockets reduce drag and improve speed as the animals course the ocean vastness between continents. Those flat black eyes, rolling back at the moment of impact, broadcast the shark's single-minded lethality to the world. We are terrified of sharks, thrilled by them, and yet attracted to the primal savagery they represent. The shark we barely glimpse in the murky water of a deep SCUBA dive represents the darkest umbra of a world we can never completely illuminate. But this kind of hyperbolic "Shark Week Science" is correct for only the smallest fraction of sharks. Most sharks through history had these basic elements (powerful teeth and a predator's diet) but ranked fairly low on the terror scale, usually grubbing in the muck for small prey.[29]

Sharks burst on the evolutionary scene 418 million years ago with a hot new innovation—a jaw—that they shared with the ancestors of bony fish. Their skeletons were made of cartilage, lighter and more flexible than bone. A specialized organ called the ampullae of Lorenzini in the head of sharks senses weak electric fields using a set of specialized ciliated cells.[30] In lab tests, sharks have been able to find hidden prey fish solely on the basis of the electrical signals the prey emits and have shown exactly the same behavior when the prey's electric signal is simulated with electrodes.[31]

All these things help define sharks. But most importantly, there are the teeth: serrated tools absolutely perfect for the task of eating in a world where the other sharpest bites were from the beaks of cephalopods. When sharks evolved 418 million years ago or so, there were beaks and claws and probosci and rasps: but nothing else had teeth. Those teeth were an amazing invention, and they gave sharks a crucial edge.

How to grow sharp

It was the invention of sharp, replaceable teeth that gave sharks their market share, their biological brand in the new world of predators that evolved in the dawn sea. Innovations are key to the success of new groups—whether it is Larry Page and Sergey Brin creating PageRank, a new way to rank the importance of web pages for Google,[32] or a tooth in a world of more and more armored animals. The innovation of the shark tooth is still paying off after 418 million years.

Sharks continuously grow their teeth—a new set every 7–10 days—and discard the old ones like dull razors.[33] And they are supremely sharp, featuring a cutting edge just one thousandth of an inch wide. They're among the sharpest natural structures on Earth.[34] Even a fossil shark tooth can cut you if you are not careful with it. These fearsome weapons have carved themselves a long and bloody legacy.

But how does a shark grow something this sharp? When humans make something sharp, we build it thin, and then pound on it or sand it down until the thin edge is thinner still. Biological structures can't be pounded or sanded this way. They must be grown already sharp, by cells and tissues that are usually far better at growing things that are soft. Hard, sharp structures were a new idea 418 million years ago. And the sharks hit on a design that is a tiny miracle of cellular engineering.

Shark teeth start deep: they start their lives deep in the mouth, beginning as a low hard ridge in the soft tissue of the throat. Ridge after ridge grow in succession and move up toward the mouth like waves rolling onto shore.[35] Individual teeth first form as an amorphous lump of tissue. Next, the tooth heightens through a thin line of cells at the crest of the ridge throwing up a narrow fan of fibers. Those thin fibers define the sharp edge of the tooth, and keeping the fibers in a dense narrow line rather than a clump is what makes the difference between a truly sharp tooth and a blunt one. Once they grow in their precise regiments, the tiny gaps between fibers are then filled in with a bone-like material, accreting slowly into nature's finest knives.[36] Finally, an even harder layer of sealing "enameloid" is applied to the cutting surfaces for extra strength. Slender structures tend to be fragile, so serration is produced by introducing undulations in the narrow ridges of fibers, thereby strengthening the ridge without thickening it. By now this row of teeth has moved into the mouth, ranking behind the those currently in use, ready for deployment if the teeth in front should crack on bone, stone, or steel.

Shark teeth 418 million years ago were very sharp but only an eighth of an inch long.[37] The sharks that made these teeth are harder to see in the fossil record, and our first glimpse of an early shark body comes from a 9-inch specimen 409 million years ago.[38] The treasure trove of shark body fossils is 370 million years old, preserved almost whole (along with their stomach contents) when ancient proto-sharks called *Cladoselache* were snatching up primitive fish using rounded grab-and-pull teeth.[39] The basic shark body plan was

A goblin shark's snout. Photographed at the Shinagawa Aquarium in Japan. Photograph by Hungarian Snow.

present but unrefined. *Cladoselache* looked like a nerdy prepubescent shark, combining a skinny body with oversized fins. This was just an early model.

A few sharks continue to look like this today. The goblin shark inhabits the deep sea, and has a huge, elongated "nose" above a scraggle-toothed mouth. Its underslung jaws with ice-pick teeth are on an elastic ligament like a rubber band. The band is stretched tight with the mouth in its retracted position, where it is held in place until prey draw near. When released, the mouth shoots forward to grab soft-bodied bottom-dwelling prey and retracts just as rapidly back into the skull.[40]

The frilled shark also has been called a living fossil because of its sinuous long body and ancient jaw structures. Though later work seems to put it in a more modern shark group, it shows how older sharks may have lived, with a

Rows of needle-like teeth on the frilled shark, used for catching soft-bodied prey like squid Photographed at the Shimonoseki Marine Science Museum, Kaikyoukan Marine Aquarium, Japan. Photograph by User:OpenCage.

6-foot eel-like body and needle-like teeth that can strike quickly and secure small, quick prey.[41]

Are sharks living fossils?

The sharks of modern seas are not much like those that first appeared 400 million years ago. Unlike the horseshoe crab, the basic body plan of sharks has not been stable over evolutionary time. Instead, the evolution of sharks from predators of bottom-dwelling invertebrates to the killing machines of modern seas is a story of extinction and refinement.

Two hundred and fifty million years ago, Earth experienced the most devastating mass die-off ever recorded. Dubbed the Permian-Triassic Extinc-

Restoration of fossil shark *Carcharodon megalodon*. Reconstruction by Bashford Dean in 1909 at the American Museum of Natural History, reported to be somewhat oversized.

tion,[42] it wiped out a staggering 96% of marine species.[43] Rapidly shifting ecological or climate changes could be responsible, and scientists have put forth theories ranging from impacts to massive volcanic events.[44] Early sharks died, but they gave rise to a small group called the Neoselachii that is the ancestor of modern sharks. They came of age in a black and empty ocean that took 5–10 million years to recover.[45] Conditions were brutal; prey were scarce. Yet the Neoselachii survived, using more powerful bodies and ever-improving dentition. They diversified, grew in size, and eventually evolved the modern sharks we thrill to today.

For example, the first great-white-style sharks of order Lamniformes came into being about 65 million years ago. They perfected their dentition and changed their mouth structures to push their jaws out.[46] While attacking, those hinges would open like flower petals to expose the ferocious teeth inside. The awe-inspiring *Carcharodon megalodon* grew nearly 40 feet long with a body mass equal to eight elephants (77 tons).[47] Its jaws, up to 6 feet across, could deliver more than 40,000 pounds of force. The biggest of its 276 serrated teeth were 6.5 inches long.[48] Megalodon probably fed on large baleen whales and evolved about when they did, about 20–30 million years ago.[49] It packed all of modern shark evolution into one body—fast, powerful, and predaceous on species that were themselves extreme in size. They disappeared in the middle of the Ice Ages 2 million years ago, for reasons so far unknown.

Some sharks once thought to retain very primitive features, like the frilled shark, are now known to be advanced families that have re-evolved the body plans of their deep ancestors. Even shark teeth have evolved from having a single enameloid layer to having three.[50] Unlike horseshoe crabs and the nautilus, sharks also remain supremely successful in the modern oceans. These evolutionary advances are no less than those made by mammals in their climb up the rungs of ecological importance, from small beginnings and marsupial ancestors, to become the dominant large animals on land.

So, are sharks legitimate living fossils? Their basic structure and body plan hasn't changed in 409 million years. The way they grow and replace teeth, and the electrical senses we cannot easily duplicate mechanically, defined them long ago and still do. Probably their greatest consistency is in their steady predatory presence in the oceans since nearly the dawn of the age of animals. Since even before the land held much more than millipedes, during the heyday of trilobites and ammonoids, during the first age of the ocean when nothing else had teeth, the sharks patrolled. As the very continents have moved, the sharks have swum their margins, searching for prey in seas alternating between abundance and extinction-caused scarcity. Living fossils? Living wonders.

Passing through history

Living fossils are more similar than their disparate anatomies would suggest: ancient organisms clinging fanatically to their ecological niches, succeeding through a handful of highly refined biological features. Certainly, speciation and adaptation in these groups have occurred over the eons. But what has changed pales in comparison to what has not. Few species manage to exist for more than a few million years. For a body plan to endure that timespan a hundred times over, fundamentally unchanged, through apocalyptic extinction events and the rise of human civilization, is astonishing—but the proof is right in front of our eyes.

Despite the contrary popular conception, evolution does not lead to progress. If anything, evolution rewards short-term success and is devoid of long-term planning. If a species successfully produces the next generation, then evolution gives it at least a passing grade. In this view, why shouldn't a successful body plan succeed for a long time? Two answers emerge: first, environments change. Second, coevolution with competitors or prey favors continuous innovation. For living fossils, long-term success seems to go hand in hand with a retreat from change. Up in the shallows, waves of mass extinctions have passed across the oceans—five so far, though humans are currently engineering the sixth.[51] These game changers reset the flow of life and cataclysmically changed the environments in which most marine species live. But some environments were less perturbed. Deep ocean environments are among the world's most stable—cold, deep, quiet, and enormous. They may simply be stable enough and big enough for some ancient forms like the coelacanth and the nautilus to survive.

And then there is the coevolutionary race among species. Much of the evolutionary play is a script that depends on the action of different players—predators consume prey, competitors exclude the weakest. When one player evolves a successful new strategy, it imposes pressure on the others to likewise evolve. Some will possess the genetic toolbox to respond, and these will continue to thrive and to change. Some will not have the toolbox, either through bad luck, small populations, or lack of genetic variation that fuels evolutionary change. Again, living fossils are likely to be those that for some reason have fallen into a way of life where few of these coevolutionary races are run.

The coelacanth and the nautilus occupy ecological niches that isolate and protect them. Two species of coelacanth are currently known, in tiny pockets separated by thousands of miles of water.[52] There are only four species of horseshoe crab, confined to the Atlantic and East Asian coasts. Perhaps six species of nautilus cruise the Indian and Pacific oceans, restricted to deep region in the tropics.[53] At the same time, and despite aggressive fishing by human beings, more than 400 species of shark ply Earth's oceans. From docile reef dwellers the size of cats to the nightmarish and diamond-rare goblin shark, these fish—for their combined evolutionary longevity and sustained diversity—might be the most successful multicellular organisms alive on our planet today.

All of which leads, we might hope, to some humility. The "living fossils," sharks excepted, are not high-profile creatures. They are too odd, too cloistered in their habitats and ecological niches to frequently intrude on human activity. It's possible that their survival is an odd accident of nature, and the world would barely miss them should they vanish. But their peculiarity is joined hand-in-hand with fragility, as niche organisms that rely heavily on their environments. The Atlantic horseshoe crab declines every year because of shoreline development on the eastern seaboard of the United States, and its decline has repercussions up and down the food chain—even as far as affecting migrating seabirds.[54] The human race is a young species, a new pack of primates who have existed for an evolutionary heartbeat. We are simply passing through history. The nautilus, the shark, the awkward coelacanth with its lobed fins—are history.

CHAPTER 3 THE SMALLEST

The ocean's smallest species have a huge impact on its chemistry and life.

More than the stars in the sky

At this very moment, there are 100 trillion bacteria on your seat. Don't reach for the disinfectant spray—those multitudes bacteria are *inside you*. The human body houses ten times more active microbes than its own living cells. It's a visceral reminder that for all of life's progress, Earth remains a microbial world.[1]

Microbes are single-celled organisms too small to see with the naked eye. The group includes bacteria, and a bacteria-like group called the Archaea that tend to live in extreme habitats, as well as some more advanced single-celled species. The microbes were some of the first living things. They burst on the world stage more than 3 billion years ago, just after the cataclysms that shaped early Earth (see Chapter 1). Their modern descendants remain the most diverse life on Earth and retain an overwhelming power to control the world's biosphere. They've explored more ways of living than any other type of creature, and today they crowd into every corner of every environment. From the jungle of your mouth to your palm's dry desert to every drop of water on the planet, microbes rule the world beneath our very noses. The chemistry of the ocean itself was crafted by bacteria, and the planet's smallest creatures play a jumbo-sized role in its maintenance.

Small packages

Microbes are numerous but tiny. Lined up end-to-end, a thousand bacteria would barely span a period on this page. Amoebae and protozoa are larger; by the seventeenth century, naturalists could peer through early microscopes and see these tiny titans wobbling through samples of pond water.[2] But their primitive lenses couldn't pick out anything smaller, and for centuries bacteria remained hidden to science. Louis Pasteur, the great French biologist, was eventually able to prove their existence while also noting that microbes caused many diseases. The "germ theory" wasn't as colorful an explanation for sickness as witchcraft or imbalanced humors, but it had the advantage of being correct. Before too long, even the most ardent skeptics yielded. Pasteur became a legend, and his theory now informs every branch of biological science.[3]

The famed Frenchman had been in his grave for a century before the technology would appear to adequately study the sea's microbial life. The Englishman Charles Darwin perceived the importance of microbes as prey for larger organisms, but confessed he was stumped by the survival of the bacteria themselves. As he wrote in 1845, "I presume that the numerous lower pelagic animals persist on [microbes], which are known to abound in the open ocean: but on what, in the clear blue water, do these [microbes] subsist?"[4]

Starting about the 1970s, our understanding of oceanic microbes was radically overhauled. Researchers led by Lawrence Pomeroy and Farooq Azam used new methods of counting microbes in seawater.[5] Their advanced techniques delivered a huge surprise, revealing not only new types of bacteria but population numbers beyond imagination. Projecting their results across the entire ocean, microbiologists faced a stunning realization: bacteria accounted for a big fraction of the ocean's biomass. Whales, fish, lobsters and every other eye-grabbing animal were fleshy icebergs floating in a vast microbial sea.

Pomeroy and colleagues projected the oceanic bacteria population at 10^{29} organisms—a 1 followed by twenty-nine 0s, an inconceivable number representing more living creatures than there are stars in the universe.[6] The number was a revelation, changing the picture of ocean ecology overnight. A column of open offshore water—an environment once thought to be unproductive—actually channels more metabolic energy through its bacterial residents than through any of its other creatures. This metabolic activity depends on diverse chemical exchanges. Some microbes consume tiny nuggets of dissolved carbon, while others dig minerals from the seawater. Multiplied by their sheer numbers, it's a staggering amount of activity. And a staggering population size. Laid end-to-end, all the ocean's bacteria would stretch thirty times around the Milky Way galaxy![7] Darwin's "clear blue water" was anything but: it was a bubbling crockpot of life.

Microbes have the potential to control the chemistry of the ocean by virtue of their numbers and efficiency. There are about as many bacteria in a quart of seawater as there are people in India, about a billion. A billion bacteria weigh only about 0.1 milligram,[8] but they can produce an outsized amount of energy if they are photosynthetic. Such is their metabolic rate that a colony of photosynthesizing bacteria the mass of 100 human beings could, under ideal conditions, produce as much energy as a nuclear power plant.[9] When the conditions are right, bacterial populations harness this metabolic energy to explode into raging blooms that shift the ocean's chemical balance.[10]

Tiny but abundant

The most abundant photosynthetic organism on Earth was something of an ugly duckling; it waited a long time to be noticed and appreciated. It showed up first as a blot of unusual pigment in water samples from the North Sea, then as a mysterious lump in an electron microscope,[11] and later as "noise" from the sophisticated instruments searching for marine microbes. The cells would eventually be named *Prochlorococcus* by oceanographer Penny Chisholm of the Massachusetts Institute of Technology and her colleagues. *Prochlorococcus* is among the smallest ocean organisms and is likely to be the most numerous worldwide, at least among the creatures we know well enough to put a name on.[12] The world's expansive seas are home to an estimated trillion-trillion individual cells of this species.[13]

This number is too big for our brains to easily process. Give this a try: if you took all the people on the planet and took all our cells (about 10 trillion per person, and there are 7 billion of us), and spread them out in the ocean, you would get about 70 billion trillion. It would take 15 planet Earths like this to equal the *Prochlorococcus* figure. So numerous are *Prochlorococcus* cells that their photosynthesis process and oxygen production support a good fraction of the rest of Earth's life. Oceanographers aren't much for grand pronouncements, but they estimate 10% of Earth's atmospheric oxygen has come from *Procholorococcus*.[14]

Once these prodigious little engines were recognized, oceanographers started to find them everywhere. They live in so many places, except the polar oceans, that they have diverged, genetically, into what Chisholm calls a "federation" of cells adapted to live in very different parts of the ocean. And just a few decades back, we had no clue they existed.

Prochlorococcus hide in plain sight by dint of sheer size—or rather, the lack of it. Measuring a mere 600 nanometers (each nanometer is one billionth of a meter and one thousandth of a micron), two or three times smaller than most bacteria in culture bottles, they fade into the background even when you're looking for microbes. Like electrons orbiting an atom, they seem to be everywhere while occupying no space at all. An average protein molecule is about 5 nanometers across,[15] so *Prochlorococcus* cells can barely fit 100 protein molecules end-to-end inside them. Space inside the cell is so valuable that even DNA gets compacted, stripped down to just 1,700 crucial genes.[16]

There are smaller genomes in nature, but only a few. A tiny parasitic bacterium exists on primate genitals, boasting only 600 genes and complaining endlessly about "the view from down here."[17] But such parasites need hosts from which to sponge a living; their numbers never grow too large. *Prochlorococcus* makes food out of sunlight and is common enough in the wild ocean to be considered the true winner of the evolutionary race for abundance. How it does this with so few genes, and does it better than the larger genomes of other species, is not yet known.

Not picky eaters

Darwin's dilemma was that all the bacteria he saw in the sea needed fuel: something that could feed a thousand trillion trillion cells. *Prochlorococcus* are photosynthetic, assembling food from sunlight and CO_2. But the smallest, simplest bacteria don't have that option. Dubbed *heterotrophs*, they're unable to use sunlight to knit CO_2 molecules together into bigger molecules, such as sugar (an ability that defines the plant-like *autotrophs*). Heterotrophic bacteria must feed on big organic molecules and break them down, usually burning oxygen in the process (as do larger heterotrophs like us mammals). They are amazingly efficient at this and almost totally indiscriminate in their tastes. They can eat anything below a certain size, starting with simple sugars. Large molecules—such as proteins and lipids—are chopped up like sushi rolls, separated into their component amino acids, and devoured. Many bacteria specialize in consuming complex oils, and some of these helped scrub the Gulf of Mexico following the Deep Horizon oil platform explosion of 2010.[18] Bottom-feeding bacteria are even capable of absorbing snippets of foreign DNA through their cell membranes. They may strip it down, gulping down atoms of phosphorous and carbon. Or they may steal the DNA outright for its information content, crudely pasting new traits into their own genomes.[19]

This utter lack of choosiness makes heterotrophic bacteria the world's greatest garbage crew. And it explains how so many bacterial cells can persist in the oceans. Every microscopic fecal pellet dropped by every planktonic crustacean is swiftly surrounded by colonizing bacteria. Sensing a new lode of precious materials, they kick their metabolisms into high gear. The pellet dissolves, breaking into smaller and smaller particles. The smallest bits, microscopic chains of carbon molecules by this point, are loosely classified as dissolved organic carbon (DOC). DOC forms strong bonds with dissolved nutrients and is a good food source, so it forms a high-efficiency slop that allows microbes to inhale calories at a prodigious rate.[20] With 700 billion tons of dissolved matter in the ocean—more than every land animal and plant massed together[21]—DOC represents the world's most eye-popping excess of food outside of a Las Vegas buffet.

The microbial loop

Suppose you are a vegan at a typical American barbecue cookout. You might be hungry, but you're surrounded by meat and dairy: food you can't eat. Large marine animals face the same dilemma. The vast amount of food tied up in dissolved organic carbon—a big portion of the ocean's whole biomass—can't be consumed by the whales, sharks, fish, or tiny copepods. Bacteria small enough to feed on DOC are themselves too small to be eaten by most predators. Instead the industrious microbes provide every creature in the ocean with an enormous indirect benefit: by acting as the ocean's recycler.

Because heterotrophic microbes will eat anything, biological "waste" is usually recycled. Bacteria consume the tiniest organic molecules, inhaling every last bit of energy and biomass, endlessly and rapidly replicating. When they die, their tiny bodies become DOC again. Many bacterial populations take 7 or fewer days to double.[22] The amount of biomass cycled through this microscopic bank is astounding.

The total bacterial biomass in the oceans is about 11 billion tons.[23] If every microbe reproduced every 7 days, they would produce about 11 million tons a minute. We humans take about 99 million tons of seafood from the ocean every year, a rate that we seem incapable of increasing. So, the global production of fish that humans use all year has the same organic content as that produced by bacteria in the sea in 90 minutes.

If this growth continued forever, the oceans would quickly fill with bacteria instead of water. But there are organisms that can quickly consume bacteria, culling their numbers like lions stalking in the Serengeti. These "lions" have manes of flagella, and though they lack claws or teeth, they are fierce predators. Every liter of seawater is a world of a billion microbes, and every

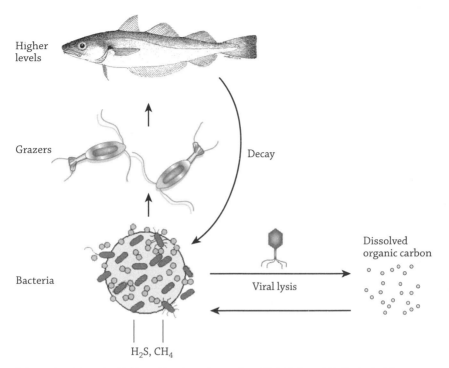

Higher
levels

Grazers

Decay

Bacteria

Viral lysis

Dissolved
organic carbon

H_2S, CH_4

Schematic of the microbial loop/virus loop. Image from Chris Kellogg, U.S. Geological Survey.

drop is its own hunting ground.[24] The predators are single-celled protists, like ameobae or *Paramecium,* and they can chew down the standing stock of bacteria as fast as it grows. That's a very good thing for the rest of the denizens in the oceans, which are spared not only a glut of bacteria but also a dearth of diversity. Protists channel this food and energy upward, into the rest of the ocean's food chain. The smallest creatures in the sea act as a universal currency for the largest and build a crucial foundation for the ocean's amazing biomachine.

It's hard to overemphasize the scale of microbial production in the open water. Years after his famous discoveries about the hidden world of marine microbes, Lawrence Pomeroy wrote, "Earth's ocean is a sea of microbes; without them it would be a very different place, less hospitable for all life. Indeed, without the activity of these organisms, the cycles of nature would very quickly come to a halt."[25] The tuna you see in the grocery store couldn't exist without the microbial productivity sustaining the open oceans, funneling nutrients to larger life through predation. A tuna in the "clear blue water" looks large and powerful, but it's sustained by trillions of invisible creatures.

Zombies of the sea

Bacteria have a lot of metabolic tricks, often including the ability to become dormant when food is lacking or the balance of nutrients is wrong.[26] Bacteria don't die of starvation, they sleep through it. They do not die of old age, they simply divide and reconstitute their DNA to be young again. Bacteria rarely die in the passive sense; they're killed.

Death often comes as described above: in the form of snarling protists, cells bigger than most bacteria and festooned with a fur of cilia or a set of whiplike flagellae. Patrolling through the water, they lack mouths to snatch and swallow their prey. Instead, each bacterium is collided with and held fast by sticky mucus. It sinks, slowly and still alive, into the protist's body, where it is dismantled in a soup of digestive enzymes.

But it could be worse.

Death also comes from below, from tiny viruses much smaller than bacteria. Ocean viruses are hugely abundant, ten times more abundant that their bacterial prey.[27] Barely meeting the definition of life as a self-replicating biological entity[28] and only the size of large protein molecules, these viruses have little motile power on their own. They are complex protein bottles with a nasty surprise inside. The viral genome—its small thread of DNA-based information—is poised to take over a bacterial victim.

Like their counterparts that cause head colds in humans, these viruses function by attaching themselves to a cell and injecting their viral DNA into the host. Viral genes use the cell's normal machinery to do what genes do everywhere: provide the instructions for making proteins. In this case, the proteins are viral proteins—proteins that kidnap the cell's metabolism and use it to make more viruses. The cell doesn't benefit and will no longer divide. Instead, it descends into living death as a slave factory for its viral controllers.

Quickly, the cell is forced to make the outer coatings of more viruses in vast numbers. Then the viral DNA is copied, coiled up, and packaged inside these simple sleeves like a plastic snake in a fake can of nuts. The hapless bacterium—now a virus factory—becomes a swollen, corrupted orb full to bursting with maggoty young. Eventually the ravaged cell can take no more, and it simply explodes. New viruses pour into the open water and disperse, already hunting for the next generation of prey.

Viral victims spew more than plagues into the ocean. The metabolic machinery of the former bacterium, battered and exhausted by the invaders, dissolves into the surrounding water like any other dead remains. Proteins, lipids, carbohydrates, and a host of rarer nutrients are then picked up by tiny heterotrophic scavengers. The remains become dissolved organic matter, feeding the microbial loop and connecting the discarded dead with the rest of the sea.[29]

This is the smallest predator-prey cycle in the world: viruses stalking bacteria, bursting them, and building up the dissolved food supply. And when scientists began looking into it and calculated how much biomass was involved, they were astounded: up to 30% of the ocean's biomass was cycling daily (yes, *daily*) around this tiny, invisible predator-prey loop.[30] Before the 1980s, both this system and its principal players were totally unknown.

Kill the winner

Bacteria are incredibly diverse: perhaps a billion bacterial species inhabit Earth, with a good fraction of them in the ocean.[31] But for all the myriad of species that ever lived, died, surged, and faded in the ocean, none ever achieved permanent dominance. Even a superabundant genus like *Prochlorococcus* is not very dominant ecologically; it exists alongside thousands of distant cousins. To this day, it's rare to see a single species or colony crowd out the others. When we do, it is a signal of an ocean out of balance, and the explosion of microbes is disastrous: beaches are closed, seafood grows toxic, and even the air near such a toxic bloom becomes deadly (see Chapter 11).[32]

What hidden mechanism keeps the ocean from filling up with one super-bacterium? Why does explosive diversity seem to be an intrinsic microbial trait? One answer might lie in the evolutionary agility of the smallest predators of the bacteria, the viruses that search them out and burst them. Their extremely rapid life cycle and genetic manipulation has made them the world's fastest-evolving predators.

A large population of bacteria, or any similar microbe, is a big prey base for viruses, and so the most successful microbes are also the biggest targets. Viruses can't infiltrate their targets without special surface proteins that allow them to enter their victim's cells. These proteins tend to be specific to one or a few types of microbes. A virus that evolves new surface proteins to target a large microbe population has a bright future. It has an enormous potential prey base and a huge potential growth rate. The cycle has all the elements of classic natural selection: mutation, predation, and tremendous reproductive rewards for the winners.[33] And this cycle has a potentially huge effect on microbial diversity, too.

Oceanographers have given this cycle a name: "kill the winner." Imagine an enormous bloom of microbes in the sea, clouding the water for miles with tiny bacterial bodies that are individually invisible. The bacteria in the bloom succeed by dint of some amazing adaptation to ocean conditions. But their very success makes them a target. Viruses mutate, and some of those mutations allow them to attack the huge hoard of bacterial prey, infiltrating the bloom. Fresh viruses exploded from a dead bacterium need only travel a few microns to the next bacterium to start the next killing cycle. Therefore, the theory goes, large microbial populations will inevitably come under intense viral attack, and the bloom will inevitably shrink. But only to a point. As countless host cells die, the bacterial population becomes so thin that the viral infection is no longer efficient. The viruses are still primed to kill, but they can't move under their own power and have to bump into their prey. When the formally explosively abundant bacterium species becomes rare, its death rate drops, and its population stabilizes.

Kill the winner is a natural safety valve for success—a blind mechanism keeping any single microbe from taking over the environment. A microbe that is too successful will be pruned down to a smaller population size. We have seen it at work: on occasion, large single-celled microbes called coccolithophores bloom in the ocean. They grow so explosively that their blooms are literally visible from space. Using overhead satellites, microbiologists can observe the subsequent viral attack. From the edges, gaps and holes start to appear in the solid mass of teeming life. Within a week, the coccolithophores are decimated, their blooms shredded like high clouds on a windy day.[34]

The constant war of kill the winner generates an arms race between microbes and viruses, with adaptations and counteradaptations. Some infected coccolithophore cells initiate a cell-suicide process called apoptosis, killing themselves, throwing themselves on the proverbial grenade, to keep the virus from copying itself.[35] The exploded cell dies but saves its daughter cells, copies cloned from itself in previous cell divisions. But evolution is happening in the virus too: more advanced versions of the virus contain a protein that inhibits apoptosis, keeping the host cell from killing itself. The most insidious may even take over this "suicide" machinery to further boost viral production.[36] Still another escape strategy is exhibited by the single-celled coccolithophore *Emiliania huxleyi*. When under viral attack, these cells convert themselves to mobile escape pods with only half their DNA.[37] Scenarios of assault, defense, and counterattack play out across the oceans and across the eons to create an amazing diversity of microscopic life.

Single-celled organisms and viruses are the fastest evolvers on our planet,[38] and "kill the winner" is Earth's single most high-stakes game. It's been played in every drop of ocean, every day, for probably the past 3 billion years. Because it keeps any one microbial species from becoming too successful, too dominant, it may be partly responsible for the balance of diversity in the sea.

A few hints from the past suggest dire consequences when the microbial world is disrupted. One example was the mass extinction of 250 million years ago that wiped out 96% of all species on Earth: massive volcanism across the globe created a huge bolus of CO_2 in the atmosphere. Massive global warming ensued, and the bathwater seas were awash in microbial blooms. The wholesale disruption of the ocean's food chains, its ecological systems of checks and balances, caused an eruption of microbes that scoured the oceans of oxygen. The resurgence of normal ocean ecosystems was derailed for 5 million years.[39]

This might sound familiar: marine biologists today are concerned about rising CO_2 levels and warming seas.[40] Evidence from the past and parallel ecological studies of our current oceans point to a dangerous time in the ocean's future if, once again, microbes come to rule the seas.

CHAPTER 4 THE DEEPEST

High pressure and low food supply
make for a difficult daily life.

On June 6, 1930, two men climbed into a hollow steel sphere, sealed themselves inside, and dropped headlong into the Atlantic Ocean. A few hours later, they returned to the blinding Bermuda sunlight, forever changed by what they'd seen. Naturalist William Beebe and an engineer named Otis Barton had descended into the blue-black depths in what would be the deepest dive in human history. Their vessel, a bathysphere, measured less than 5 feet across and offered only a porthole of fused quartz to look through. A single steel cable and a rubber air hose ran up to the surface, narrowly distinguishing the bathysphere from a tomb. They reached 800 feet—a monumental achievement that placed their names in history books—and came back because "some mental warning . . . spelled *bottom* for this trip."[1] But the bathysphere had proven itself, despite just the smallest of leaks, and Beebe and Barton kept pushing their dives deeper. Every new dive was a life-or-death experiment, allowing them to refine both their equipment and their methods. On August 15, 1934, Beebe and Barton reached a depth of 3,028 feet—inspiring the title of Beebe's wildly popular account, *Half Mile Down*.

William Beebe was more than a scientist. He had an explorer's soul and a poet's pen. He unshipped them both to introduce the lumbering surface dwellers of planet Earth to the delicate ballet and the toothy terrors of the world's darkest deepest places. What Beebe saw beneath the waves was completely alien to the human experience: pitch-black water, monstrous-looking creatures, delicate jellies, and the unending flicker of tiny lights in the darkness. A half mile down under tons of barometric pressure, Beebe felt more than any-

William Beebe and Otis Barton standing next to the bathysphere.

thing else the weight of absolute loneliness: the knowledge that he was past the human world, viewing sights never before seen:

> nowhere have I felt so completely isolated as in this bathysphere, in the blackness of ocean's depths . . . we seemed like unborn embryos with unnumbered geological epochs to come before we should emerge to play our little parts in the unimportant shifts and changes of a few moments in human history.[2]

The dark unnerves us. Whether it's in the space below the stairs or beyond the campfire's edge, people are nervous about surprises that lurk unseen. The deepest reaches of the sea really are akin to another planet, subject to conditions unimaginable on Earth's surface. Crushing pressure, deep cold, and eternal darkness rule the world's basement.

Beebe immediately understood what he had accomplished. The bathysphere was a simple craft—a tiny helpless sphere with nothing but an open vial of soda lime to absorb CO_2 from their breath—and yet it transported the adventurous pair to another world, entirely beyond human experience. One simple communication from 600 feet, sent by copper telephone wires up the air hose, conveys the weight on Beebe's mind, like so many tons of salt water overhead: "Only dead men have sunk below this."[3]

Deep-sea creatures are hugely varied, displaying a fascinating (and occasionally nightmarish) array of forms and adaptations that mold their lives to the pressures and temperatures around them. In the abyssal deep where the Sun never rises, razor-toothed fish hunt by deception, and giant worms dine on a soup of boiling chemicals. There is a quiet flow of simple beauty—delicate bodies built of jelly spinning their gossamer lives unseen and alone. And always, as Beebe observed, there is the flashing green glow of bioluminescence—light created by living creatures. Jellies sport glowing aquamarine racing stripes, and fish dangle lit lures that entice prey with a calculated spasticity. Helpless creatures set off their own fireworks when attacked, drawing even bigger predators in desperate life-or-death gambits.

Though Beebe's journeys were pioneering, and his discoveries fundamental, the science was simple: observe, report, survive. His scientific assets were a camera, a good photographer, and an encyclopedic knowledge of deep-sea species brought up by nets and dredges.[4] In the 80 years since Beebe's dive, with a wealth of modern technology, we've come to understand a great deal more about the lives of deep creatures that the great naturalist couldn't begin to know.

Landlords of the deep

The most valuable food source in the ocean—the Sun—is completely absent in the depths. Photosynthesis is no longer an option where light is missing. And so no deep-sea mother ever demands "Eat your vegetables," because there are none.[5] Instead, in the abyssal dark, every species is a killer, a scavenger, or a landlord.

The landlords live in the neighborhood of Earth's cracked floor. Boiling water gushes from continental wounds, agitated by the planet's tremendous heat and laced with sulfurous poisons. The cracks are called hydrothermal vents, or more colorfully, black smokers.

The first hydrothermal vents were investigated in 1976, stunning the expedition scientists with their dense and diverse populations of new species.[6]

The slow delivery of food means that life generally lives on thin rations in the deep sea, and any one species usually has low abundance. Vents did not appear at first to be different: under a modern submersible's floodlights, the vents appear as black clouds spewing from spires of accumulated sediment.[7] The chemicals are poisonous to most terrestrial life and smell like rotten eggs.[8] Yet hydrothermal vent communities were lush life in the deepest ocean darkness, hidden oases in the world's most barren desert. Where was the food supply for this bonanza?

The food and the rotten egg smell both come from a simple chemical, hydrogen sulfide. For all its toxicity, the molecule's sulfur bonds practically crackle with energy. Living in and around hydrothermal vents are bacteria that have mastered *chemosynthesis:* the conversion of the chemical energy of hydrogen sulfide into raw cellular energy. Chemosynthetic bacteria break the noxious sulfide molecules apart and use the resulting release of chemical energy to fuel microbial growth.[9] That mastery lets them build new cells and power their metabolism by cracking the chemical energy bubbling from Earth's crust.

Hydrothermal vents house a fascinating collection of bacteria living in a very odd place. But these deep communities consist of much more, because a unique set of animals has evolved to exploit this microbial abundance. Worms and bivalves and shrimp all live off the microbes that live off the sulfur. Shannon Johnson, an ecologist studying deep-sea organisms off the California coast, puts it succinctly: in the blackened deep, "Everything has to live with or off of bacteria to survive."[10] But they do not merely consume the bacteria—they rent out living space for them.

Lips or leaves?

The microbes' most visible beneficiary—the abyss's most prosperous landlord—is *Riftia pachyptila*, the giant tube worm. Nicknamed "living lipstick," the worms' white bodies are capped with vivid red plumes like folded flower buds. Worms take root near vents, building pale chitinous tubes to house themselves and growing up to 5 feet in length.[11] The plumes resemble a pursed pair of lips, but *Riftia* has no mouth—these worms lack even a gut. Instead, they have organs near their plumes called trophosomes: fleshy sacs containing so many microbes they account for a good chunk of the worm's body mass. The worms are sustained entirely by their resident bacteria, which process hydrogen sulfide from the hot water and pipe the excess product to the tube worm's systems.[12]

A new species of crab clusters around hydrothermal vents in the Southern Ocean near Antarctica. From Rogers, A. D., P. A. Tyler, D. P. Connelly, J. T. Copley, R. James, et al. 2012. "The discovery of new deep-sea hydrothermal vent communities in the southern ocean and implications for biogeography." *PLOS Biology* 10(1):e1001234. doi:10.1371/journal.pbio.1001234.

The lip-like structures are the deep-sea equivalents of leaves gathering sunlight. Absorbing huge quantities of hydrogen sulfide, CO_2, and oxygen into tissue of the red plume rich with capillaries, they bind the molecules to a familiar protein: a hemoglobin like that found in our own blood. From the worm's plume, the hemoglobin transports these molecules to the bacteria in the trophosome.[13] Sulfides, oxygen, and CO_2 are the ideal metabolic fuel to keep the trophosome factories running and the tenants content.[14]

Because *Riftia* tube worms have no mouth and no gut, they depend entirely on microbes for their food supply. But young worms are not born with on-board microbes, instead acquiring them during the larval stage. For decades, biologists thought the larval tube worms just swallowed the bacteria they needed, using a tiny mouth-like aperture that only the larvae had.[15] Newer research reveals the truth: bacteria invade the young worm through its skin; later, the worm forms the trophosome to hold them. With the right microbes aboard, juvenile worms jettison their digestive tracts and live by digesting the excess bacteria growing in the trophosome.[16]

Like many rent-seeking arrangements, this is a fantastic deal for the land-lords. Most deep-dwelling organisms grow and reproduce slowly, stunted by cold and deprivation. *Riftia* tube worms defy the trend, growing at astounding speed. Cameras on the Pacific floor have recorded tube worms colonizing a new site; they settled, reproduced, and grew to a thicket nearly 5 feet high in less than 2 years.[17] They are among the fastest-growing marine invertebrates on the planet, standing in stark contrast to some of their own relatives. Other tube worms eschew hot vents for quiet deep-sea "cold seeps": cold, slow hydrothermal vents with similar chemical output. They grow to equally impressive size, but take more than 200 years to do so.[18] The extreme conditions at hydrothermal vents are hazardous, but they catalyze spectacular growth at Earth's most inhospitable sites. More than 500 new species have been identified at hydrothermal vents in the few decades since their discovery.[19] Many more will likely be found in the Southern Ocean near Antarctica, where the first expeditions have already uncovered a zoo of new species living at the hottest places in the coldest sea.[20]

A whale oasis

Hydrothermal vents aren't the only deep-sea oases bursting with unexpected life. Life blooms wherever resources exist, and not all bottom-dwelling communities are based on chemosynthesis. Some are fed by the riotous productivity of the sunlit layers thousands of feet above. Surface-water detritus, flakes of tissue, pieces of algae, and fecal matter, drift down in spiraling white flakes, solitary and slow, whimsically dubbed "marine snow." The descent can take weeks, and most of the material is consumed high in the water column before it reaches the bottom. Bacteria are a major beneficiary of this trickling snowfall (see Chapter 3's discussion of dissolved organic carbon), but snow is also a valuable food source for scavengers when it reaches the bottom. Still, living on marine snow is a thin diet for larger animals and does not generally support abundant species in dense communities.[21]

There is, however, an occasional bonanza for denizens of the deep, like winning the lottery for worms. A dead whale is an oasis on the sea floor.

Whales don't fall like snow.[22] Instead, they deliver huge lumps of meat to the ocean's bottom, falling rapidly in one place and all at once. The largest whales are open-ocean swimmers, so they tend to die in cold, deep water. When they strike the muddy bottom with a soft gurgling thump, the scavengers of the deep go to work.[23]

An armada of hagfish, sharks, and squid arrives, somehow rapidly locating the carcass in the sea's vast wasteland—a mysterious process akin to finding the only open coffee shop at midnight at a giant airport. A multi-ton whale carcass is stripped of most of its soft tissue within a few months.[24] True bottom-dwelling organisms—mollusks, worms, crustaceans, and other simpler life forms—are largely absent, while fast swimmers tear through the carcass. This is the mobile scavenger phase, the first of three generalized stages of decomposition of whale falls.[25]

Once the meat is gone, it's time for the next meal. Then the enrichment opportunist stage begins, when communities of bottom-dwelling scavengers colonize the carcass. Crustaceans and polychaete worms make up this phase, but a second wave of mobile predators also arrives to set up shop, ignoring the carcass to feed on the sessile scavengers. A community of scavengers and their predators feeds on the whale for months or years. Eventually, only the leviathan's skeleton remains.

During the third and final period, what was a grim (if busy) boneyard transforms into an unexpectedly flourishing oasis. The whale's flesh may be long gone, but its bones are filled with valuable oil. Bacteria take the lead, dissolving the bones while feasting on the lipids inside. These microbes aren't Archaea, but they employ similar chemosynthetic processes, using dissolved sulfates to digest the whale bones.[26]

Whale-fall oases may fill a special and heretofore-underappreciated role in the ecosystem of the sea floor; they may act as stepping-stones between larger oases. Hydrothermal vents are ephemeral. They start and then stop, sometimes over the space of mere years. They may be separated by thousands of miles; only Earth's fickle crust can say. Because some of the same species are seen at vents all over the world, some brave souls must have traversed the desert to populate new sites. Whale falls may fill in the gap: tiny isolated pockets of vent-like life peppering the gulfs between the more robust vent communities. Scientists estimate that more than a half-million whale falls exist in the planet's oceans at any given time, deposited every few miles along the biggest migration routes.[27] These are outposts, way stations in the darkness where sojourners may settle and reproduce. Their descendents will push farther into the desert, long after the original colonizers are dead.[28]

If organisms in the deep ocean have always been concentrated at vents and falls, their presence raises a troubling question. How has whaling affected deep-sea ecosystems? Human beings have, by conservative estimates, destroyed three-quarters of the planet's great whales. Extrapolating from that hideous figure, the ocean floor must receive only a small fraction of the whale falls it used to. Oases, rare now, may have been far more common in the past. By hunting the great whales, we've spent centuries starving and fundamentally altering ecosystems that relied on their dead bodies.[29]

All-female blind zombie worm: *Osedax*

In February 2002, a research scientist at the Monterey Bay Aquarium Research Institute stumbled on a whale fall in a deep chasm offshore. Piloting the remotely operated submersible *Tiburon*, Robert Vrijenhoek sighted a gray whale skeleton on a ledge nearly 2 miles down. Though perfectly and magnificently complete, the bones didn't glow the typical eggshell white under the *Tiburon*'s lights. They were mossy and gray, carpeted by a mysterious red growth. Moving closer, Vrijenhoek could see the color came from hundreds of crimson filaments, swaying slowly, even though the water was still. At the touch of a metal claw, they retracted instantly into the amorphous gray. This was not vegetation. Extracting a sample to the surface, the intrepid little robot introduced science to an entirely new species: *Osedax mucofloris,* the zombie bone worm.[30]

Whale-fall communities are highly diverse; up to 200 species may populate a single oasis.[31] *Osedax* stands out among them all for its sheer eye-popping weirdness. The Latin name of *Osedax mucofloris* translates to "bone-eating snot flower," and rarely has taxonomy been so descriptive. An irregular mass of brightly colored gelatinous tissue about the size of a fingernail, *Osedax* looks like nothing so much as the contents of a tissue after a sneeze. A single, delicate stalk of flesh extends into the water column, absorbing oxygen like a single strange gill.[32]

Osedax has no mouth and no digestive tract. Like its relative the great tube worm, it relies on symbiotic bacteria to survive. But whereas a *Riftia* worm pulls sulfur compounds from the water near vents, *Osedax* extracts nutrients from whale bones. The gelatinous body conceals its most crucial adaptation: specialized tendrils on the underside. Like tiny drills, the roots of the worm bore relentlessly into bones. A potent pairing of enzymes converts seawater into a powerful acid as the tendrils branch out like tree roots, scouring out lipids and conveying them to symbiont bacteria inside the *Osedax*. The end result is a highly efficient eating machine. Colonies of *Osedax* spread across the whale bones, riddling them with tiny Swiss-cheese apertures.[33] They're much faster than bacteria alone, and before long the skeleton is broken apart: scattered across the sea floor like so much shattered marble.

The little snot flowers had one more surprise in store for zoologists, who wondered why they could never capture an adult male. As it turned out, the answer was simple: there were none to find. Adult male *Osedax* do not exist. Only females grow to true sexual maturity; males remain stunted larvae. Nearly microscopic, they're condemned to live as sperm-producing dwarves serving the vastly larger females. A typical large female holds dozens of males near her, protecting them so they can churn out sperm to fertilize eggs. But they are never fed. Males live their entire short lives off the egg yolk they were spawned with.[34]

At first, it appeared *Osedax* needed whale falls to survive. Their adaptations are so incredibly specialized that it seemed unlikely they would work well with other food sources. But scientists have convinced the worms to consume cattle and seal bones in a laboratory, and the eerie, branching tunnels typical of *Osedax* "roots" have been observed in the fossilized bones of prehistoric marine birds.[35] Recently, Greg Rouse and his colleagues dropped fish bones into the deep sea to see what life they attracted. *Osedax* appeared in these piles, showing more initiative than previously supposed.[36] The ocean's floor is an endless catacomb, and *Osedax* work tirelessly, sweeping the darkened halls.

Gas and volume in the deep sea

Away from whale falls and outside hydrothermal oases, the abyssal ocean is barren and lonely. Cold, darkness, and starvation slow life to a crawl. But it's not just the cold, the dark, the emptiness—it's the pressure. Anyone who's jumped into a pool's deep end knows the sensation: steely pneumatic fingers digging into your ears and pressing on your cheekbones. You feel the weight of every water molecule between you and the surface. Although the pressure is manageable at 12 feet of depth, an abyssal environment is something else entirely. With miles of water overhead, pressure in the deep sea can reach up to 1,000 atmospheres, or nearly 15,000 pounds of force per square inch of body surface. Even the most sophisticated submersibles must carefully hover above crush depth—the level at which the pressure will sunder their titanium hulls.

One of the biggest problems with pressure is that gas under high pressure compresses into a much smaller volume. Take the example of a diving seal. A human would take a full breath before a long plunge, but the seal does the opposite: she exhales before a long, deep dive. Pressure on her lungs increases

A Styrofoam cup that has been taken into the deep sea by the Johnson Sea-link. Image courtesy of Ross et al., NOAA OE, HBOI.

from 1 to 10 atmospheres as she drops 400 feet from the surface. Boyle's Law decrees that air will compress to a tenth its previous volume during this dive.[37] The seal's lungs obediently shrink. Because they were not initially full of air, the lungs compress essentially to a solid mass.[38] This prevents nitrogen from dissolving into the seal's blood supply—too much nitrogen in the blood causes "the bends" in human divers who descend too long and ascend too quickly. The other advantage of collapsed lungs is that they're no longer buoyant, allowing the seal to spend less energy powering down from the surface to ever deeper depths.

Styrofoam fun in the deep

By and large, scientists who work in submarines are not a happy-go-lucky kind of group. There is one whimsical tradition, though, completely in accord with Boyle's Law, that deep sea biologists are known to follow: before a dive, they'll strap a simple Styrofoam cup to the hull. The tiny sequestered gas bubbles that make Styrofoam such a good insulator surrender to Boyle's Law, shrinking the cup both dramatically and permanently.[39]

There is no particularly clever point to this ritual. It may be simply that the souvenir reminds each deep-sea explorer that they were in an extraordinarily odd and different place.

Margarine of the deep

When pressure reaches extremely high levels, it interferes with more than Styrofoam cups—it begins disrupting the very working of cells, because pressure alters how molecules react to one another. Animal cells are swaddled in an outer membrane made of lipids: hydrocarbon molecules loosely classified as fat. Normally, portals made of proteins in the membrane move nutrients and ions in and out, regulating the cell's function, feeding it, removing wastes. But when membrane lipids are subjected to extreme pressure, they thicken and harden like bacon grease congealing in a mason jar. Cell membranes become congealed, and their gates shut down. The cell is unable to get what it needs, is unable to communicate with its fellows, and unable to function properly.[40]

In response, the deepest animals have re-engineered their membrane chemistry. A different lipid is used for membranes in the deep, one that remains fluid even under enormous pressure. One of the ways deep-sea animals accomplish this is by decreasing how much saturated fat they put in their cell membranes. Saturated fat is "solid fat," made up of straight carbon chains with only single bonds between carbon atoms. This arrangement lets these molecules stack up solidly under pressure or cold temperature, like pieces of lumber stacked up in a lumber yard. Butter, meat, and dark chocolate are high in solid saturated fats and congeal easily, clogging up human arteries.[41] In contrast, unsaturated fats have one or more double bonds between adjacent carbon molecules, which introduces a kink in the chain and keeps these compounds more fluid under pressure or low temperature. These molecules are more like bent tree branches than straight lumber—they do not stack up easily.

Margarine is low in saturated fats—the straight molecules, which constitute only 10–20% of the lipid total—and has a high proportion of unsaturated fats (the bent molecules). So, it does not congeal as easily as butter. As a result of this chemistry, an abyssal membrane is made more like margarine and less like butter: abyssal membranes contain fewer saturated fats. Surface fish like salmon tip the scales at about 35% saturated fat. Three miles down, evolution under high hydrostatic pressure has fundamentally altered the same tissues: fish contain only 10% saturated fats.[42] At depth, the loose lipids are compressed by pressure but retain the right fluidity for function. And of course, the process works in reverse. Deep-sea animals brought to the surface don't behave normally, even if treated with care. Chaos ensues as their unsaturated lipids melt and proteins fail to function at low pressures. As a result, modern deep-sea biologists must collect specimens very carefully. Organisms captured at extreme depth survive poorly unless they are brought to the surface in super-pressurized containers.

But many of the most interesting creatures will not fit into the small "upmersibles" that shipboard biologists use to retrieve deep-sea specimens. Because, despite the cold and the pressure and the lack of food, over the eons of evolutionary time, a few species have skewed the calculus of survival in their favor—by simply growing larger.

Deep-sea giants

Most deep dwellers have cousins in more shallow waters, displaying differing colors, behaviors, and genetics.[43] Along the west coast of the United States, the deep sea urchin *Allocentrotus fragilis* is a pale, fragile-shelled sister species of the common, tide pool purple sea urchin, *Strongylocentrotus purpuratus*. The fragile urchin was one of the first deep-sea species to have its whole genome sequenced and compared to a shallow-water relative. It's long string of 28,000 or so genes is marked by change after change that signal the evolutionary transformations needed in the deep sea. It was not just a few genes that needed altering, but many.[44]

Some evolutionary changes adjust the basic size and growth patterns of a species. Many deep-sea animals have evolved small size, probably as a consequence of deep sea's eternal famine.[45] But some have separated themselves from their shallower cousins by paradoxically evolving in the opposite direction—a class of adaptations broadly termed deep sea gigantism. Animals of astonishing size have been dredged from the darkness; movie-monster variants of common creatures.

The giant isopod, *Bathynomus giganteus,* is one such species. Isopods are crustaceans with layered armored plates on their backs. They exist on land as well—the pill bug or potato bug, known for curling itself up into a ball is an isopod. The giant isopod is essentially a 20-pound pill bug. The largest measure about the size of a bag of Doritos corn chips, about 2 feet long from mandible to tail.[46] Twelve scrabbling legs adorned with hooks and pincers sprout from beneath a broad, armored shield. Large compound eyes are set deep in the peach-colored plates of its face; head-on, it displays a menacing glare.[47] This horror-movie beast, for all its creepiness, leads a simple life. It's a scavenger and opportunistic predator, content to gnaw on a corpse or to devour slow-moving benthic invertebrates.

The giant Isopod *Bathynomus giganteus*. From the National Oceanic and Atmospheric Administration's Ocean Explorer program. Photograph by Ryan M. Moody.

The precise factors behind deep-sea gigantism are a contentious subject among biologists.[48] The giant tube worm, thankfully, is a relatively easy case: a simple, sedentary animal force-fed huge amounts of energy by symbiotic bacteria.[49] For other deep-sea animals, gigantism occurs in cold, well-oxygenated waters—especially at the poles. More oxygen in the water may mean it is easier for creatures to supply oxygen to tissues deep in their bodies, especially for crustaceans that have relatively poor gills. But giant isopods also occur in poorly oxygenated water. This and other inconsistencies have presently shelved gigantism's "oxygen hypothesis."[50]

Two other features of the deepest sea might encourage gigantism: low temperature and environmental stability.[51] Low water temperature is associated with larger cell size and larger overall bodies, but every cold-water species should theoretically benefit. Environmental stability should also be experienced by all species in the deep sea, making long life spans a better evolutionary bet than in shallower, more volatile waters. But when it comes to betting on survival, different evolutionary tactics can succeed even in the same environment.

Some species delay sexual maturity and continue to grow late in their lives before reproducing profusely.[52] Funneling food energy into growth at the cost of delayed reproduction may yield greater success down the road if the animal is larger and can produce more young.[53] But delayed reproduction is always a gamble, because any animal can die at any moment. So some species with high mortality rates breed like guinea pigs, growing only to the minimum reproductive size before throwing themselves into parenting. The deep sea fosters both strategies. Fast reproducers colonize oases near vents or whale falls, exploiting temporary resource booms to spread their genes. Other species take advantage of the deep sea's slow-and-steady environment, living long and hoping to prosper.

Cephalopods of unusual size

For centuries, authors from Melville to Verne to Beebe romanticized the abyss. The edges of maps crawled with sea monsters: fantastic apparitions born from wonder and fear. As modern science directed its bright light into these dark places, the mysteries have faded one by one. But a demon of the Dark Ages survived, a legend never totally laid to rest. It is perhaps the only organism on the planet to inextricably bind together its own fact and fiction: the giant squid.

It is one of the ironies of the vastness of the sea that two types of behemoth cephalopods remain largely hidden from human view: *Architeuthis dux* and *Mesonychoteuthis hamiltoni,* the giant squid and colossal squid, respectively. The giant squid may grow a bit longer, though the colossal squid is broader, thicker, and much heavier.[54] Casual accounts from the nineteenth and early twentieth centuries measured dead *Architeuthis* at up to 60 feet: *Mesonychoteuthis* might reach 80. But these numbers have never been confirmed. The largest captured colossal squid was about 33 feet in total length, from the tip of the mantle to the tip of the longest tentacles.[55] Giant squid likely reach a maximum of about 40 feet: two animals of roughly this size reportedly washed up in Newfoundland in Canada in 1870 and are entered in the Smithsonian's records of *Architeuthis* from published reports.[56] A squid's body is just over half its total length (the rest is tentacles),[57] so a 30-foot specimen has a 16-foot body—about the size of a minivan. No other animal this size on Earth remains so poorly known (we think).[58]

Architeuthis prowl open ocean waters the world over. They are active predators, eating fish and other cephalopods. But they are also prey: their young are eaten by dolphins, fish, and even seabirds.[59] But sperm whales (*Physeter macrocephalus*) are better squid hunters than any species, including humans (see the Prologue). Sperm whale stomachs often hold undigested beaks the size of softballs, and their flanks are adorned with battle scars from the hooks and spinning saws of mighty squid.[60]

Recently, a band of scientists pooled samples from around the world to examine the genetics of giant squid.[61] They found two odd patterns: first, even though they had samples from all the world's oceans, the squid's genetics suggested a global species with no separate populations in different parts of the world. Second, there was little genetic diversity among giant squid. It seemed as though the global population had only recently expanded from a bottleneck in just the past few hundred thousand years. These genetic history patterns eerily resemble those of the giant squid's chief predator, the sperm whale, which also has low global population differences and low genetic diversity consistent with recent population expansion.[62]

The colossal squid masses higher than the giant squid—up to 1,100 pounds—and sports vicious hooks along its arms and tentacles. It is restricted to the remote Southern Ocean around Antarctica, and so few specimens have been recovered. The largest on record was snared by a fishing boat in the Ross Sea in February 2007. The squid was chomping on Patagonian toothfish (usually called Chilean seabass) in the boat's nets, stubbornly and foolishly refusing to abandon its fine meal. Once frozen, the squid was given to the Museum of New Zealand, where it can still be seen today.[63] Steve O'Shea of the Auckland University of Technology quipped it would make "calamari rings the size of a truck tire."[64]

Until 2004, neither giant nor colossal squid had been seen alive in its natural habitat; all the world's knowledge came from beached or floating carcasses, from an occasional animal near the surface, and from the squid beaks that accumulate in sperm whale stomachs like so much loose pocket change. But then a Japanese remote submersible captured one of the world's most elusive images: a 15-foot giant squid in its habitat, hunting 3,000 feet below the waves.

Tsunemi Kubodera of Tokyo's National Science Museum and Kyoichi Mori of the Ogasawara Whale Watching Association had spent countless hours trolling the deep ocean for giant squid with little success. When luck finally broke their way, the cameras were ready. Video from the dogged pair shows a squid in action, a single floodlight illuminating silver skin as the animal advances toward a nugget of bait. Arms extend and blossom with geometric perfection, studded with white suction cups lined with serrated teeth. There is an amazing grace to this apparition, arms swirl in a ballet of exploration of the caged bait, and then the animal moves gently away as if spooked. Grainy pictures do no justice to the scale of this apparition; those arms could enfold a family sedan. Yet the animal was never aggressive except in the businesslike way of predators. Popular images of giant squid show them cherry red with rage (as happens when at the surface, fighting for their lives) and thrashing their tentacles like whips. Kubodera and Mori captured *Architeuthis* in its element, breathtaking and utterly in control.[65]

Where are the truly huge specimens, like those Captain Nemo fought on the deck of his *Nautilus*? It is possible that exceptional individuals exist, somewhere in the ancient blackness beneath the planet's seas, hidden under ice caps, or among the volcanoes of the vast Pacific. If a creature grew beyond the strength of the largest whales, no mysterious squid beaks would appear as gastric evidence of the existence of a third giant squid species. If it lived its life in the hidden vastness of the deep sea, avoiding submarines, we'd have no way to know about it. Imagination will always tempt us more than reality: we'll always draw monsters on the margins of maps.

The magic light: Bioluminescence

We know about species because we drag them up to the surface and name them. But in the deep sea, these species know one another as well. They do not have name tags like you might at a tedious high school reunion. Instead many of them are bedecked with lights.

Imagine yourself a tiny, helpless fish in the limitless darkness of the deep sea. The blue-black water has neither ceiling nor floor: it is like dark moonless sky stretched above and below. Yet the endless night is not peaceful. You are forever being watched by hundreds of eyes anxious to grasp just a few slivers of light. Predators lie concealed in the dark all around, gnashing needle teeth in untold numbers. And at any moment, a tiny trickle of sunlight leaking in from up above might betray you.

But there is other light—like the moonless sky has stars. A constant flickering of blue and green surrounds you, faint and furtive spikes of light that could mean anything from a fresh meal to a gruesome death. The deep sea is the only ecosystem on Earth—barring deep caves where only fungi grow— where the main source of light is not the Sun, but an organic protein.

Luciferases, or photoproteins, generate light by splitting high-energy molecules—producing photons instead of metabolic energy.[66] Some fish have luciferase genes and array these glowing proteins inside a small pit in the skin, a specialized light-producing organ called the photophore. Most fish secrete their own luminous compounds, but some grow sacs packed with symbiotic light-producing microbes.

Bioluminescence is the sea's most important tactical adaptation. Deployed in careful patterns on the underbelly, photophores of some fish match their output to the faint light coming from above. Thus a fish below them cannot detect their passage as they cruise above, giving them stealth as predators or prey.[67] Simple plankton make a great deal of light noise, blaring photons at the slightest disturbance and filling the deep sea with inane visual chatter.[68] This chatter may actually serve a purpose—experiments have shown that when a shrimp consumes some types of plankton,[69] the plankton shoot off bioluminescent sparks. Predatory fish rush in like a SWAT team, attracted by the alarm call, devouring the shrimp and ignoring the tiny plankton. Marine biologist Steve Haddock and his co-workers have recently documented as many as seven different defensive roles for bioluminescence in deep-sea fish.[70] There are offensive roles as well: bright lights can stun or confuse prey, attract them to dangling lures held in front of monstrous jaws, or serve as high-beam headlights to find morsels drifting in the water column.

Perhaps the most famous of these innovators is the anglerfish, which hunts with a long fleshy lure. The generic "anglerfish" describes an entire family of spectacularly ugly animals. Lacking dorsal fins, the anglers have moved the spines that would typically form these fins forward to a spot just above the eyes. The first spine is thickened and greatly elongated into a protruding digit and crowned with an irregular bulb of tissue called an esca: a bioluminescent lure.[71] The lure's spongy tissue is thoroughly inhabited by hard working, light-producing microbes. They cause the esca to glow enticingly in the dark water, as the host fish sells the illusion. Like a seasoned bass fisherman, the angler makes its lure unbearably enticing. Twitching, bobbing, turning in loops like a frantic frolicking worm, a glowing blob becomes an irresistible target. Little predatory fish, dwarfed by the angler, approach the esca and strike with all the force they can muster. They vanish instantly, without sound and barely a perceptible movement: sucked into the huge mouth and impaled on a flash of needle teeth. Every anglerfish species has its own unique esca, some longer than the fish's body, all luminescent.[72] How the anglerfish detects nearby prey isn't clear—their eyes are small and poor. Some have sug-

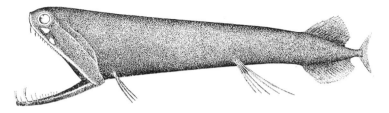

Stoplight loosejaw, *Malacosteus niger*. Published in Goode, B. G., and T. H. Bean. 1896. *Oceanic Ichthyology*. Special Bulletin 2. Washington, D.C.: Smithsonian Institution, plate 37.

gested the killing reflex is triggered by the slightest touch on the lure. What has become clear is that anglerfish will attack fish of almost any size. The record is a 4.5-inch anglerfish with a foot-long rattail fish in its mouth. Both predator and prey were dead when they were caught floating off the shore of Papua New Guinea.[73]

Most luminescence is blue-green to match the deep sea's weak sunlight. But the loosejaw group of deep-sea dragonfish (family Stomiidae) project a unique hue.[74] Large and powerful photophores just beneath their eyes beam red light through the water. They accomplish this through a unique fluorescent protein in some species and a red-brown filter over the photophore in others.[75]

Red is an unusual color in the deep sea. Seawater absorbs red light and more easily transmits blue, and so most of the bioluminescence in the sea is in a far-reaching blue-green hue. The predators and prey of loosejaws have eyes particularly sensitive to this blue and green light, having evolved beneath a mile of seawater.

The loosejaws are a rare exception, specially evolved to see the red light that they themselves produce. The protein that fish use to capture light in their eyes, opsin, is specially modified in the loosejaws. At position 261, a unique mutation changes an amino acid that is critical to the way the protein absorbs light. As a result, this protein in loosejaws absorbs much more red light than do other deep-sea fish, and they can see the prey reflected in their own unique spotlight.[76] Most deep-sea creatures can only flicker their lights—quick on/off bursts lest they are discovered and devoured by predators. In a dark world of killers, bright signals illuminate food but also summon death. The loosejaws—small and feeble compared to surface predators—have beaten this game, seeing red without being seen, prowling the abyss with crimson impunity.

All of the lights

The true character of the deepest ocean isn't embodied by a stalking auto-mobile-sized squid, or hundreds of 6-foot tube worms sprouting like weeds around a black smoker. When we imagine these things, we miss the vast bulk of the deep ocean. We typically think of water as clear, of light as omnipresent, of big things moving through empty spaces. The nature of the real abyss was captured by William Beebe. As he sat in his tiny globe, in the night of the sea, what stuck with him was not any of the fantastic predatory fish he'd seen—it was the lights. They bloomed against the darkness, twinkling and pulsing and filling the bathysphere's tiny quartz porthole. All around him the lights spoke to one another in an unread language, telling a story of life and death, and the deception of desperate predators. Imagine these animals not as images from textbooks but as they would appear in their own world. Deprived of all light and cloaked in dark scales, they know one another only by the flickering of bioluminescence and the hint of a black silhouette.

As the first man to visit the deepest sea, Beebe felt an enormous responsibility. Having seen what no living person had ever seen, he was compelled to describe it. He understood that he had been to another world beyond terrestrial Earth. Writing 30 years before the first spacewalk, Beebe prophetically captured the abyss:

the only other place comparable to these marvelous nether regions must surely be naked space itself, out far beyond atmosphere, between the stars, where sunlight has no grip upon the dust and rubbish of planetary air, where the blackness of space, the shining planets, comets, suns, and stars must really be closely akin to the world of life as it appears to the eyes of an awed human being, in the open ocean, one half mile down.[77]

CHAPTER 5 THE SHALLOWEST

*For marine life perched above high tide, survival is about balancing
the danger above against the danger below.*

The streamlined sea urchin

For 20 years, the stone wall at Kaka'ako Park in Hawaii has been a shield
against the Pacific Ocean. Boulders weighing thousands of pounds fit together
like giant bricks laid down over miles of shore, tightly knit without a drop of
mortar. Children leap from one to another as they chase each other down
the shoreline. Tourists stroll along the waterfront. Families set out food on
the picnic tables. And all the while, the waves keep rolling in. They march in
ranks, pounding the black stones with the strength of ages.

Along the waterline, lower down than the children are usually allowed to
play, hard purple domes the size of silver dollars cling to the rocks. These are
shingle urchins, genus *Colobocentrotus*. They lack the menacing sharp spines
of typical urchins. But the spines remain there, in a fashion—protruding
from the animal's base like a skirt, rounded and dull as Popsicle sticks. Atop
the animal's body, the spines are very short and shaped like tiny mushroom
caps that fit together like the stones on the Kaka'ako wall. And like the wall
they are shields against the sea.

Shingle urchins persist largely by dint of two adaptations. First, the
unusual topside spines provide a smooth low profile to reduce the drag of
water across the animal. Second, the sea urchin uses its unusually strong
feet to grip the stone. Hundreds of tube feet adorn the urchin's underside,
using tiny vacuum chambers to create suction. Each is just a spaghetti-wide
filament, but together, the tube feet keep a death-grip on the rock. All sea
urchins have tube feet—the shingle urchin's are simply stronger.

These adaptations make sense intuitively, but how well do they each function in practice? A simple experiment can answer this question. By attaching

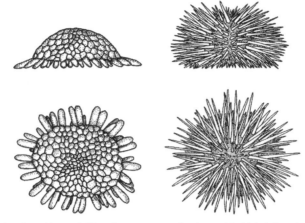

A shingle urchin from Hawaii (left) and a more normal spiny urchin (right). Drawing by Freya Sommer. Reproduced/adapted with permission from Denny, M., and B. Gaylord. 1996. "Why the urchin lost its spines: Hydrodynamic forces and survivorship in three echinoids." *Journal of Experimental Biology* 199(3):717–729. Doi: http://jeb.biologists.org/content/199/3/717.full.pdf.

normal urchin spines to a shingle urchin, we can test whether the tube feet are strong enough to hold the animal on the rocks even with extra drag. If so, tube feet are more important than maintaining minimal water drag. If not, the smooth shingled dome is the more crucial adaptation.

Brad Gallien, then a graduate student at the University of Hawaii, hollowed out the globe of sharp spines from a typical urchin and affixed them like a helmet to the smooth top side of a shingle urchin. He reintroduced the chimera into the maelstrom of the Hawaiian surf. Observation showed that the added spines hugely increased the pummeling that the urchins took, but they were still able to hold on. Clearly, the powerful tube feet (also used for feeding and movement) are more important to the urchin's safety than its unique spines.[1]

There is another element at work here: positioning. Waves break on the urchin's bulwark, but water still penetrates beneath the shielding spines to keep the animal moist. Urchins do not have lungs or gills, so they must absorb oxygen by means of simple diffusion through wet surface tissues.

Like other sea urchins, the shingle urchin can survive without water for a short time, but eventually it will die. But—oddly for a marine species—it will also perish if submerged for too long. Oxygen diffuses more easily in air than water, so it's easier for the urchin to breathe if it's moist and stationed above the waterline. Shingle urchins have considerably less breathing capacity than other urchins, possibly due to their dense armor. If trapped under water for even a few days, they drown.

It's strange to imagine, a sea urchin fearing the water. But in fact, every intertidal organism faces a similar dilemma. The high intertidal may be dry and hot, but the low intertidal is a savage jungle of predators and competitors. Whether on a sandy beach or rocky shore, everything that lives between high and low tide sits balanced between the twin poles of desiccation and danger. As a result, intertidal life organizes itself parallel to the water line: laid down in horizontal bands like parallel grooves on an old record. When the tide recedes, the intertidal's changing variables are laid as bare as the substrate—expressed in the distribution of creatures along the coast.

These bands, or zones, can be seen almost universally on the planet's shorelines. As far back as the 1930s, T. A. and Ann Stephenson were drawing them.[2] From shoreline to shoreline, all across the globe, the husband-and-wife team found similar horizontal patterns. It was clear to the Stephensons that a logical system was at work, based on slight changes in environment from one stratum to the next. "Zonation results from . . . gradients," they asserted.[3] In their minds, the gradients were both biological and physical—but there could be no doubt that intertidal creatures arranged themselves based on proximity to the water.

At any point along the shore, conditions will be ideal for something to live. As you march from the higher parts to the lowest zones, two conditions are in constant flux. One variable is the environmental hazard posed by the Sun and dry air. These are marine organisms, after all, and any time they spend exposed above the waterline taxes their systems.

At the other end of the spectrum, the water is a dangerous place. It's filled with big, strong, open-water creatures. Competition and predation—biological hazards—mount as the environmental stresses of Sun and wind wane.[4] Put simply, the farther organisms live above the waterline, the more environmental pressure dominates. The farther they live below the waterline, the more biological pressure dominates. Every point on the substrate is a unique combination of the two dangers—a mix of oil and vinegar that changes with tidal height. Every creature on the shore boasts adaptations specialized to its exact place along the shore: to its own precise combination of environmental and biological stressors.

To live farthest from the ocean

The highest intertidal region is the splash zone. Here, the substrate is seldom wet; the only water comes from large waves, sending spray higher than the tides will ever reach. The rocks are dusted with colonies of algae, lichens, and tough little snails.[5] With so few organisms able to tolerate the dry conditions, they've got a lot of space to themselves. It's rare to see one species crowd out another. True predators are rarer still—restricted to cracks in the rocks that shelter them from the heat and exposure. The end result is the highest, driest, and least-diverse zone in the intertidal. The "high life" is safe, but to live here is to specialize in discomfort.

Over decades, the rocks themselves are shaped by the feeding action of intertidal snails and other mollusks. Chitons in particular have teeth on their tongue-like rasp (called a radula) that are hardened with magnetite, allowing them to erode rock. Some protected tropical shorelines have been transformed by these tiny rasps: notches in the shorelines of the mushroom-shaped rock islands in Palau are the most magical effect they produce.[6]

The periwinkle snails *Littorina* are among the highest-living of all marine snails, picking spots on the rocks that are splashed by wave action maybe once or twice during each 2-week tidal cycle. Small globes not much larger than pebbles taper to a sharp point at the top of the shell. They live by rasping thin algae and debris off the stone with their tongues. They are named for the littoral zone, the region of any coastline that stretches from the high-water mark to the point of total inundation. Cast them into a tidepool or a bucket of water, and they immediately begin a long slow climb out of it. They will not drown as shingle urchins do, but their preferences are clear. To the snail, water represents a lethal hazard rather than a survival need.

Which adaptations allow this particular animal to survive out of water? There are several, all premised on a Fremen-like efficiency with water. Like most sea creatures who live above the tides, *Littorina* require a certain amount of fluid to keep their gills moist. On dry days, they secrete a sticky goo that adheres them to their rocky homes using a minimum of water. If conditions are very hot, they may release some stored water and cool themselves through evaporation much as people sweat.[7] This tactic dries the snails out quickly, so tropical *Littorina* do not survive as high on the shore as do their temperate cousins. They are also much lighter in color and more spherical so that they can hold more fluid for cooling.[8]

But the adaptations that allow for the success of these snails can also be used against them. Periwinkle snails aren't found only on coastal rocks; variants of the same animal, in the genus *Nassarius,* spend their lives in the salt marshes of the U.S. East Coast. When the tides come in and the muddy marsh basins fill with water, snails slide up the long blue-green grasses in retreat. Again, this is a delicate balance. Hazards abound both above and below. Predators in the water are matched by birds in the sky and crabs in the mud. In a typical intertidal compromise, the snails crawl up the marsh grass but stop short of the top. Nonetheless, *Nassarius* can be seen from time to time at the very tips of the marsh grass. Their weight bends the slender stalks, letting the breeze metronome them to and fro. Before long, the snails are picked off and devoured by birds.[9]

Why do snails engage in such suicidal acts of daring? The answer is a tiny parasitic worm called *Gynaecotyla* that infects the periwinkles and modifies their behavior. Like an insidious sci-fi monster, the parasite takes up residence in the animal's brain and begins to erode its fear of heights. Before long, what once seemed a healthy height no longer satisfies. So the snail moves higher, embracing the end. For the parasite has needs of its own; it must escape the snail to move on to the next phase of its own development. This occurs when the snail itself is eaten; only then can the worm complete its development in the predator and reproduce.

Where cooperation rules

The salt marshes that harbor the confused periwinkle snails are massive habitats, dominating the coastal zones of the American eastern seaboard. Streams wind their way through muddy lowlands, cutting through ground so soft that anyone stepping off the trail finds himself knee-deep in stinking black mud. The tide ebbs and flows through these channels, flooding the flat land twice a day. Like a rocky coastline, the marsh environment is divided into zones based primarily on subtle differences in elevation. At the lowest levels, called the mud flats, fine silty soil is inundated with water several times per day. Above this empire of mud is the first of the salt marsh plants. In the well-studied marshes of New England, you would find thick stands of cordgrass (*Spartina alterniflora*).[10] But the broad swaths of simple grass belie the biological complexity that lies beneath. And it takes a different kind of interaction among species to keep this habitat stable.

The roots of the marsh grass are not deep, but they must hold the silty earth together in the face of the turning tides or a raging storm. The marshes catch and tame the waves that crash in from across the Atlantic. They collect the sediment washing downstream from the inland hills and keep the loose silty soil from spilling into the sea. As a result, these plants form the foundation of the coast. They anchor much of the eastward edge of North America. But for the grasses to do their work, they need an ecosystem where the whole is more than the sum of its meager parts. They need help from other organisms to survive.

Ecologists call this phenomenon *facilitation*. The dense root networks safeguarding the coastline also shelter small animals. The plants keep the water from flowing too quickly, providing a safe environment for sedentary mollusks. In the New England salt marshes, this role is filled by the ribbed marsh mussel (*Geukensia demissa*). The ribbed mussel is a bivalve, like clams and scallops: their two-piece shells are joined by powerful hinge ligaments. And their main business is to filter large volumes of water during high tide, and extract their food from it. Their fertilizing waste is a great boon to vegetation, falling as it does directly into the root network. In a salt marsh where resources are scarce, this kind of relationship helps out the whole ecosystem.[11] How do we know the mussels make a difference? Mark Bertness of Brown University conducted the experiment needed to be sure: he took mussels away from the system, watching the marsh grasses dwindle and the soil to lose essential nutrients.[12]

The hidden marine mussels and the tough marsh plants do not look like a team, but they function as one, each gaining a critical advantage and providing critical help. In this way, they literally create new land. Other species also help: fiddler crabs scrape a scavengers' living out of the rich new muck, burrowing into it as a defense against both predators and desiccation. In the process, they add oxygen and nutrients to the marsh mud, enhancing cordgrass growth.

Organisms in the marsh, through facilitation and cooperation, are environmental architects, profoundly manipulating the structure and composition of their homes.[13] Like coral reefs and mangrove forests, they help to build and maintain the habitats that shelter them.

Mangrove forests

In many tropical regions, the land's edge is defined not by beaches or rocky shores or salt marshes but by mangroves forests. Webs of gnarled wood may stretch for miles, turning blue lagoons into impassable thickets. Dozens of hard knobbly roots sink into soft sediment, supporting the mangroves as they rise out of the mire. The environmental challenges faced by these trees, like the grasses of salt marshes, are so daunting that few other plants could survive. The salt, the wave action, the intense sunshine, and the suffocating lack of oxygen in the soil all work against successful plant life. In fact, mangroves are the only trees on the planet that can survive with roots immersed in salt water.[14]

Despite these hardships, they've become massively successful all over the tropics. A severe environmental handicap has become a competitive edge. How does this happen? The first answer lies in the stilt-like roots.[15] The illusion of a floating forest is made possible by the massive root structures under water. Mangrove roots, heavier and harder than those of most plants, bring much-needed physical structure to a constantly shifting environment. Roots dive down vertically into the sediment like pier pilings, disrupting water flow with a dense maze of inch-wide wooden staves. Nutritious sediment from coastal rivers gets trapped before it reaches the ocean, and entire ecosystems appear in the shelter that these roots provide.[16]

Oysters, sponges, sea squirts, and algae form a thick carpet over the roots. Shrimp and small fish flit by through the arched naves of this magnificent underwater cathedral. As in the marshes, these animals' waste products enrich their surroundings. The teeming life sheltered in the mangrove cradles help keep the forest well fed.[17]

A problem of salt and oxygen

Salt, like so many nutrients, is necessary for life but lethal in large doses. Organisms living in a marine environment are deluged with salt and have evolved ways to shed it from their systems. Sea turtles secrete excess salt through their tear ducts, and a crocodile's famous tears are saltier than ocean water.[18] Salt marsh cordgrasses have salt glands that excrete salt on the leaf surface to be washed or blown away. Some mangroves have similar adaptations: the white mangrove (family Combretaceae) takes its name from the white salt deposits on its leaves. Two secreting glands sit at the base of each leaf, and they release excess salt along with small amounts of water. Over time the leaf becomes encrusted with salt crystals, causing the tree to shine white in the tropical Sun.

Other species of mangrove take a vastly different approach, refusing to absorb salt in the first place. Red mangroves (family Rhizophoraceae) are typically found in the most waterlogged coastal environments. Their roots are different from those of their white cousins: they are a natural desalination plant. Rather than absorbing ocean water through sloppy diffusion, the roots pump it through special filters before supplying the rest of the tree.[19] This allows the mangrove to select which molecules it wishes to pass on to the rest of the plant and which to filter out. With this system in place, up to 99% of the tree's salt intake is shed at the roots. A red mangrove absorbs no more salt than a tree in your back yard.[20]

Mangrove plants face another persistent hardship: suffocation. A land plant cannot adequately absorb oxygen if its roots are totally submerged. What's more, the sediments at the bottom of a mangrove forest are a soup of active microbes. In the course of breaking down waste matter, those bacteria consume a great deal of oxygen—so much that mangrove sediment is nearly devoid of oxygen. For most plants, the combination of water inundation and low-oxygen soil would be lethal.

As before, different mangrove species have their own peculiar adaptations to this challenge. The red Rhizophoraceae live in the deepest water and steepest tides of any mangrove, and their root structures hoist them farther above the water than their cousins. Their bark is studded with tiny structures called lenticels, which are essentially pores for taking in oxygen.[21] Red mangroves also grow "cork warts" on leaves—a specialized tissue of hollow cells that lets oxygen diffuse into the plant, where it then flows into open spaces in the stems and travels down to the roots.[22]

Other species extend snorkels vertically into the air. Black mangroves (family Avicenniaceae) grow specialized roots called pneumatophores up out of the ground instead of down below it, reaching heights of several feet. Pneumatophores are covered with lenticels and another type of specialized tissue called aerenchyma: light and airy tissue, short on mass and long on surface area to pull in oxygen. Pneumatophores are another adaptation that, for the vast majority of terrestrial plants, would be a waste of biomass and therefore disadvantageous. In the upside-down world of a mangrove forest, they make perfect sense.

A fish out of water

Fish are drawn to mangrove forests by the lush diversity between the trees' knuckled roots. Many species of juvenile reef fish begin their lives among the mangroves, traveling to offshore coral reefs only when they've grown large.[23] There are predators as well, lurking in the shadowy labyrinth. Small fish can hide from predators in the narrowest gaps, but there is one other escape direction: up. In a mostly submerged world, real estate above the water line

The mudskipper *Periophthalmus modestus*. Photograph by Webridge.

is a precious refuge from predation. It's for this reason that some fish have learned to escape the water and take a breather on land.

Mudskippers, a group of fish in the Goby family, thrive in intertidal habitats around the world by surviving long stretches out of water. Mudskippers are skinny mottled-brown fish with giant froggy eyes atop their heads. They are fast little creatures, catching their prey with bursts of lightning speed and wolfing them down between tiny needle teeth. Pectoral fins have grown strong enough to support the fish's own weight even without water's buoyancy. They can ambulate slowly in a motion called crutching, scooting themselves along with those fins.[24] A skipping motion is also possible, wherein the animal launches itself in the air wth a flip of its powerful tail. Jumps of up to 2 feet have been recorded by mudskippers skipping between tidepools; no small feat for an animal barely 4 inches long.[25]

Mudskippers can move out of water, but how can they be so active without oxygen? As it turns out, the little fish have evolved to breathe air. Their mouths, throats, and gills are lined with oxygen-absorbing mucus membranes that serve as inefficient gills so long as they're wet.[26] That lack of efficiency explains the mudskipper's enormous mouth, throat, and jaws: it needs more surface area to breathe. Any time it ventures out of water, it holds an air bubble in its mouth like a SCUBA diver's tank.[27]

Mudskippers abide in underwater burrows like bunkers in the mud, providing protection from high tides and high heat. At the bunker's rear sits a pocket of air, scooped out during construction and replenished from time to time by the skipper. When salty, low-oxygen runoff inundates their burrows, mudskippers nurse their air bubbles while staying safe below.[28]

Middle earth—the mid-intertidal

Littorina snails and mangroves live mostly out of water, where environmental stress is highest and their physiologies are under constant assault. But just a few feet lower on the shore lies a densely populated stratum of life representing a more even balance of stresses. This is the mid-intertidal zone, exposed at low tide but submerged at high tide.

On temperate shorelines, barnacles mark this region: small crustaceans cemented head-down on the rocks, filtering food from passing waves with frond-like feet.[29] Barnacles hunker in thick armored shells that protect them from wave impact and dehydration. At the same time, they need regular immersion to eat and breathe. Larval barnacles are carried by the waves themselves and must be careful not to take root too high, lest they doom themselves to an early grave.

Monsters on the shore

As with every intertidal zone, lower climes are filled with monsters. Sea stars and whelk snails live by devouring helpless, immobile barnacles—but at the same time, they are damaged by the Sun's desiccation if they venture too high. These predators can climb up the rocks at high tide, inch by arduous inch, but must flee before the low tide maroons them too high up the rock face. These predators face a conundrum: they often take so long to eat a meal that, if starting at high tide when the rocks are covered in comfortable ocean water, they may still be at it when the tide turns and the sea retreats.[30]

The snails called whelks are predators on animals with shells—often the barnacles that are so common on mid-intertidal rocks. But barnacles are boxed in hard bony plates that keep most predators out. Whelks thwart their prey's defenses using a unique mouth structure simply dubbed a drill. The drill bores a round hole in a hard shell, and pierces the animal inside. The whelk then sucks out its prey's fluids like a mosquito drawing blood. But this vicious technique takes a long time, both to penetrate the shell and liquefy

Starfish *Pisaster* feeding on a mussel. Photograph by Linda Fink.

the prey with digestive enzymes. Often the whelk must abandon a half-finished meal to retreat with the falling tide.

Starfish also eat slowly—they center themselves over clusters of barnacles and smother their prey by extruding their balloon-like stomachs. A quicker meal comes by pulling the barnacles off the rocks one by one. Starfish glue themselves down securely using the tube feet at the ends of their arms, and they employ other tube feet near the centrally placed mouth to maintain a death grip on their prey. The thick muscles in their arms contract and pull upward, ripping the barnacles off the substrate.[31] The animals are then jammed into the starfish's mouth, where digestion begins, letting the predator amble back down toward the tideline.

How high on the rock barnacles live is determined by access to water and the nutrients it carries. A barnacle too far up the shore will dry out and die before it grows big enough to reproduce. But the lower bound is set by predators; a barnacle too far down grows rapidly but soon becomes a juicy target for predators. The end result is a band of mottled white and gray on coasts across the world: barnacles by the thousands, squeezed bulwark-to-bulwark in the stratum between the infernal Sun above and the hungry predators below.

Mussel beach

Directly below the rough, spackled barnacles on many temperature shores is a dark band of purple and black. These are mussel beds: bivalve mollusks packed as densely as a medieval graveyard and covered with a moist sheen like the eggs of a giant alien insect. They're sessile feeders like the barnacles and resemble them in many ways. Yet for mussels, the calculus of biology and environment is just a little different. Mussels have a weaker and more open shell—in fact, the "at rest" position for their hinge muscle allows the shell to stand ajar—and lose moisture constantly. Thus they can't live as high on the rocks as barnacles but must occupy a lower zone. They also need more protection from the battering waves. Mussels glue themselves down to the rock with a thread called a byssus. The thread is first made as a viscous, protein-type glue in a small gland. The mussel attaches one end of a strand to the rock, and the other to itself. When the glue sets, the thread holds the animal in place along with hundreds of its fellows.[32] But mussels can slowly move, like climbers high up a vertical rock face, by securing a new byssus thread connection to the rock face ahead of it and breaking the old one behind.

A shrieking gull can pry open mussels to scoop out the animals inside, or just swallow them shell and all. Oystercatchers swing their thick beaks to hammer shells open. Despite all these aerial threats, marine predators remain mussels' greatest menace. Sea stars lumber up from below during high tide, pluck an unlucky mollusk from the rocks, and flee at a glacial pace back to safety. The mussels closest to water get nabbed first, establishing the purple band's lowest boundary. It's rare to see any mussels below this point unless they are lucky enough to grow too big for starfish to eat.

One of history's most famous ecology experiments might have been the simplest. Robert Treat Paine III traveled to a shoreline in the Pacific Northwest of the United States, where mussels form a broad slate-blue band in the middle intertidal. There, he began pulling all the starfish he could find off the rocks and tossing them unceremoniously into the nearest cove. Month after month and year after year (Paine has been at this for some 49 years), the starfish have been stripped. The experiment was to see whether the predatory starfish were active enough to control where the mussels could live. The answer was a resounding Yes: the mussels, free of the starfish menace, migrated lower down on the shoreline than any of their fellows elsewhere. Paine's simple experiment makes it safe for mussels to be a lower intertidal species, proving the lower boundary of the mussel zone is set by predator pressure and not by physical stress.[33]

Carpet of life on the lower shores

Point Pinos is roaring. Waves detonate on the shore, heaving white foam into the air and dusting the rocks with salty spray. Low-growing strawflowers with brilliant purple blossoms eke out a living on thin sandy soil; cypress trees keep hunchbacked vigils on the bluffs above. Vast tracts of kelp roll in the surf offshore like battle debris. This spit of rocks on the California coast, at the southernmost point of the Monterey Bay, is one of the world's most spectacular coastlines.

Pacific swells pound the high intertidal, but beneath the surface their effects are very different. In the low intertidal, waves churn rather than striking—pulling water back and forth across the substrate.

Kelp plants called *Laminaria* are perfectly built for this constant churn. Brown fronds protrude from long ropy stalks that twist like lithe dancers in the surf. If placed in a higher zone, they'd be stripped bare by the waves or baked by the Sun. The easier physical conditions of the lower intertidal have created a swath of softer, wetter, and more mobile organisms than seen above. They're also more numerous, as the low intertidal has easy access to the ocean's great stores of nutrients. With total immersion, rolling waves, and stable salinity, marine plants grow taller and healthier. This extra productivity at the food web's base gets magnified exponentially throughout the ecosystem.

The lower intertidal is what most of us would consider the ocean. Submerged nearly all the time, it's visible only during the lowest tides. At Point Pinos, the rock substrate is thoroughly hidden under a carpet of vegetation. A shag carpet of red, green, and brown glistens on the rocks like Atlantis dredged from the depths. Physical hazards are minimal most of the time: organisms don't need special protection against the Sun or desiccation. They need only endure them for a spell before the water covers them again. On the other side of the coin, the biology is dense. The lower intertidal pulls resources from both land and sea, and those riches attract a lot of life.

Stinging in the rain

The big sea anemones of the Pacific Northwest (*Anthopleura xanthogrammica*) have adapted to feed off the welfare of the waves. Lying sedentary along crevices and channels, the tentacled predators wait for waves to deliver snails, small crabs, mussels, and all manner of other prey. With an arsenal of stinging cells and a gaping toothless maw, anemones look like tiny replicas of the Great Sarlacc in *Return of the Jedi*. For like the Sarlacc, an anemone swallows

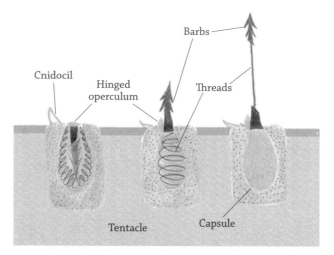

Barbs

Cnidocil

Hinged
operculum

Threads

Tentacle

Capsule

Nematocysts have a trigger called a cnidocil, which opens a hinged operculum to release a barb attached to a hollow thread. This feeding mechanism is found in cnidarians and is the reason jellyfish sting.

everything that touches its maw. Only the hardest, most unappetizing remnants are disgorged later.

The stinging cells of an anemone are small football-shaped structures called nematocysts. Coiled inside is a sharp barb connected to a hollow cord, and a well of powerful toxin. The apparatus is linked to a simple contact trigger that, when activated, launches the barb out toward the disturbance. Tethered like a harpoon, it pierces its target and delivers a tiny flood of venom through the cord. Small creatures are paralyzed in seconds, though to large animals like humans, anemones are mostly harmless.

Yet some of the anemone's relatives produce hugely painful stings. The Portugese man-of-war raises welts on human skin that burn like a necklace of wasp stings. Most anemones do not have these abilities, but their nematocysts are nevertheless powerful enough to subdue their food. Powerful enough to be a valuable weapon, even in the clutches of other creatures.

Without a shell, they need other protection

Nudibranchs are marine snails without shells. Usually carnivorous and always beautiful, there's a nudibranch for every color of the rainbow. Adorned with frond-like appendages that make them look like tropical flowers, they scour the shallow rocks of almost every shoreline in search of prey. Their soft bodies need protection, and some species find it in the form of toxins. Bright colors and antler-like head ornaments signal the nudibranch's toxicity to the rest of the ocean, but the poison is actually stolen from their prey—anemones included.

Anemones cover themselves with a layer of mucus as a form of self-protection, specially formulated to avoid triggering its own stinging cells. When a nudibranch attacks, its first move is to rub itself against the anemone—slathering on mucus while carefully avoiding the mouth and tentacles. Once swaddled in goop, the nudibranch is free to feed with impunity. Tearing off chunks of anemone and wolfing them down, it digests the flesh while somehow sparing the nematocysts. Passed through the sea slug's digestive system, the stinging cells are collected and repatriated to the slug's skin. The autonomous little harpoons don't realize where they are and continue to function as normal, stinging any predators that might attack the sea slug. Through this remarkable adaptation, a soft unarmed creature makes itself dangerous prey.

Making a life at the edge of the sea

Creatures of the low intertidal rarely can survive higher up. They are ill equipped for the physical stresses of sunlight and air. They languish, dry up, and die. Instead, they are armored to endure predation and competition—biological threats. In this way, the low intertidal represents one extreme of the great coastal gradient: a region where environmental challenges give way almost entirely to biological challenges.

Likewise, high intertidal species, moved a few feet down, would be devoured by much larger and more capable ocean-going predators or tougher competitors. Their adaptations to survive desiccation would be useless. Below the high intertidal line, these organisms are simply out of their league—the competitors and predators are too big, too fast, and too powerful. These are the same gradients the Doctors Stephenson recognized decades ago—and they rule intertidal life all over the world.

The story of the intertidal is a war story, a struggle between two opposing forces. As with most disputes, the border between the two is the heart of the struggle. Biological pressure and environmental pressure are the warring sides, and they duke it out along the terrain from high tide to low tide marks. The rules are simple: the farther you live from the water, the more dangerous your environment is. The closer you live to it, the more dangerous your neighbors are. Would you like to live on the Siberian tundra, or in the heart of Rio's most violent *favela*? This fundamental tension is reflected in everything we see at Point Pinos, from the wind-bowed cypress to the otters trolling through kelp a hundred yards offshore.

The striated bands of the intertidal are battle lines—physical representations of the Stephensons' classic gradient, laid out on every coast of our ocean-bound planet. Each species has struck a compromise with different aspects of nature, and the specific terms dictate where they live. The terms of the compromise are laid as bare as the Kaka'ako seawall, codified in the adaptations these creatures devise. Even in the narrow firmament between land and sea, life strikes a balance.

CHAPTER 6 THE OLDEST

*Some of the most familiar species
live unusually long lives.*

Bomb carbon

Dawn breaks over a quiet Pacific atoll. A thin ribbon of beach glows with the golden color of the rising Sun. Ghost crabs skitter across the sand. Light breezes lap at the edges of a calm lagoon ringed by low islands, turquoise waters, and coconut palms.

In a flash, everything is gone.

A second sun ascends from the Pacific Ocean atop a mountain of fire. Superheated vapor roars outward from the explosion, engulfing a fleet of old warships abandoned offshore. Roiling smoke propels the cloud skyward as a thousand thunders roar across the ocean. Now a mile into the sky, the living fire molds itself into a plumed cap above a smoky stalk. An unmistakable mushroom cloud takes shape. It is 6 A.M., March 1, 1954, and the United States Army has just detonated the world's first hydrogen bomb on Bikini Atoll.

Operation Castle Bravo was the most powerful weapon in U.S. history, twice as powerful as expected and equivalent to 15 million tons of TNT. The test island was completely vaporized. Radiation flew on the wind's wings and contaminated more than 7,000 square miles of ocean and the populations of four other atolls.[1] Today, the fallout from Castle Bravo has largely dissipated. But it hasn't disappeared: the hydrogen bomb's legacy endures—in every drop of ocean water on the planet, and inside the bodies of deep-sea fish.

Carbon aging

Greg Cailliet, a fish biologist at Moss Landing Marine Labs on Monterey Bay, California, has worked for years determining the ages of fishes. Bigger fish are usually older, but beyond that simple rule it is extremely difficult to tell a 2-year-old fish from one that has lived a century or more. So when people like Greg and his students began looking closely, they discovered that many fish are far older than we ever thought.

Why is this something worth knowing? It has mostly to do with how fast a population can grow. Fish that live and die quickly tend to reproduce at a higher rate and replenish themselves faster. These are weedier species and can be fished harder—taking more for our kitchens and leaving less in the sea.[2] But information about the age of fish does more than help stock our tables: it teaches us a great deal about the ocean's secretive residents, the tempo of generations that pass beneath the waves.

How would you determine the age of a fish? Unlike human beings, they show few overt signs of aging—no gray hairs, no rheumy eyes. The answer lies, oddly enough, in the ears—or at least in the ear bones. These bones, called otoliths, grow as the fish grows. Year by year, thin rings of new bony material build up on the otolith, like rings in the trunk of a tree. Peering through a microscope at carefully polished sections of these tiny ear bones, fish biologists painstakingly count the rings. But species add rings at different rates; some yearly, others seasonally.[3] So the number of rings, on its own, isn't quite enough. An internal marker is needed; a reference point that sets the growth rings in context.

Even an accurate map, or a perfect data set, is useless without a reference point from which to start. Greg Cailliet and his students needed a starting place, too, and he found it in the radiation that burst from hydrogen bombs.

While examining Pacific rockfish in the lab, Greg and students Allen Andrews and Lisa Kerr decided to look for deposits of carbon-14 in the fish's otoliths. This unstable variant of the standard carbon atom is forged by hydrogen fusion in infant stars, but once formed, it starts to decay. Our planet's stores of carbon-14 are maintained by a balance between decay and cosmic-ray production. But human beings created a lot more of it, without even meaning to, on Bikini Atoll. Hydrogen bombs trigger a small nuclear explosion harnessing its energy to jump-start hydrogen fusion. For a split second, a star is born—and with it the unmistakable mark of carbon-14 in living things, including people, around the world.[4]

How did this material, formed in the explosion on Bikini, come to rest in the bones of deep-sea fish off the coast of California? Rockfish live in deep water, but while young, they swim close to the surface, where nutrients and prey are plentiful. Fish that were young during the age of atmospheric nuclear tests absorbed radioactivity as it spread over the world. Labeled fish larvae, with the central core of their otoliths infused with carbon-14, carried this mark deeper into the ocean as they grew and sank to their adult depths. They kept this mark, an invisible cipher, etched into their bones.[5]

Greg and his whole lab of students began to test whether carbon-14 would help solve some of the prevailing mysteries of age in deep-sea fish. They began looking at a wonderful fish called a yellow-eye rockfish. Yelloweyes (*Sebastes ruberrimus*) live in deep water off the coast of California, a few hundred to a thousand feet deep.[6] Growing up to 3 feet in length, they have been prized as a commercial species and are often called red snapper in markets on the U.S. West Coast.[7] In 2001, the species was declared severely overfished—down to 7–13% of its original numbers, because it was not growing and maturing fast enough to replace those caught by fishermen. The problem of overfishing was surprising, because yelloweye rockfish were not fished that much harder than other fish. So Greg and colleagues began looking for the reason this species seemed so sensitive to fishing pressure. The answer was age.

Once they started looking, Greg's students found that the otoliths of many yellow-eye rockfish had a core of carbon-14—meaning that these specimens had actually been alive as larvae when the bombs went off.[8] But there was another, even more interesting set of fish—the largest fish had no carbon-14 signatures. These fish had been born before the age of atomic testing, before the ocean surface was layered in new carbon-14 from fusion bombs and before Greg's students were even born.

The patterns of bomb carbon in rockfish otoliths was proof positive these fish were already older than anyone thought. This information, plus other work from colleagues who had been looking at other isotope signatures of age, showed clearly that yelloweyes often live to 100 years and beyond.[9] They continue to bear healthy young up until their deaths; no reproductive senescence is observed. Thus older, larger fish produce a lot of the eggs for the next generation,[10] and the loss of them to fishing hooks knocks down the egg production for the future. In addition, the age at reproductive maturity is also high for any fish, about 20 years,[11] and so it takes a long time for a population to replace the large, old fish so prized by the fishing industry.

The revelation that we simply did not know basic details of these fish's lives was a far larger problem than changing some numbers in a textbook. Greg notes that deep-sea fishing is a modern phenomenon; for most of human history, the fish we caught were familiar shallow-water species. But with time, we began to move farther offshore and exploit deep-sea marine resources we didn't fully understand.[12] "Some of the deep-sea fishes . . . live longer, grow slower and indeed are more vulnerable," Greg says. "They do what they can to survive, and [suddenly] they're facing another source of mortality."[13]

How could any fishery be sustainable when the animals are so slow growing and old? The solution, Greg insists, has two sides: aggressive research by scientists about what the fish could sustain as a fishery, and cautious exploitation by fishers. "Sustainability [means] you can fish somewhere forever and the fish will stick around. . . . Before you allow a fishery to exponentially increase its effort or its catch . . . you need to know the basic life history."[14]

Bowheads

It was in 1993 that an Alaskan Inupiaq Eskimo sunk his knife into a ragged old scar of a newly killed bowhead whale (*Balaena mysticetus*) and hit something solid. Beneath the leather-tough skin and the thick layer of blubber lay an impossible object—the broken-off tip of a stone harpoon—a handcrafted weapon made with ancient techniques not used for 100 years. The Native Americans had stumbled across their own ancestors' century-old handiwork, lodged in the flesh of a whale that had survived a long-ago attack and outlived its attackers. Archaeologists confirmed the weapon's origin and thus the age of the whale: this animal had been wounded more than 100 years before, battling for life in the polar seas while Napoleon III warred with Prussia.[15]

Bowhead whales are large, slow, lumbering filtering machines that sift vast quantities of small planktonic food from Arctic seas. The largest grow to a length of 50 feet and can mass 130,000 pounds (ten times the weight of a large adult elephant).[16] They boast blubber a half-yard thick and a heavy, bony skull specially adapted to break through ice floes. Bowheads are part of the family of "right whales," so named because they were so oil-rich, slow, and valuable that whalers considered them the right whales to kill.[17]

The chance discovery by Arctic natives of ancient harpoons upset the careful calculations of the world's whaling management, the International Whaling Commission, which had assumed bowheads to have a lifespan of barely 50 years. The first harpoon was carefully ignored. Subsequent discoveries of old harpoon tips[18] demanded more attention. And demanded some confirmation from methods that might measure whale ages more directly. Bomb carbon isn't useful for bowheads: their teeth are soft baleen, not the hard stones

A humpback whale breaching off the coast of Angola. These animals combine tonnage with acrobatics and adroit swimming, at least in part because of their uniquely large fins. The bumps on these fins have been mimicked in high-efficiency fan blades. See Chapter 7. Photograph by Steve Zalan.

Sailfish herd baitfish into compact balls for their high-speed harvest. The dorsal sail is used to corral the prey. Some of the fastest fish in the ocean, sailfish and their cousins, the swordfish, sport adaptations to make high-speed feeding possible. Swordfish heat their eyes and optic nerves above seawater temperatures so that they can function quickly enough during rapid feeding. See Chapter 7. Photograph by Maurizio Handler / National Geographic Creative.

The oldest known animal—4,270 years old—is a deep-sea black coral, similar to the one shown here living deep off the coast of Hawaii. Stable ocean environments with few predators or storms favor long life for animals such as these that reproduce more and more as they age and grow. See Chapter 6. Photograph by Brian J. Skerry / National Geographic Creative.

© E. Widder

A female deep-sea anglerfish (*Melanocetus johnsonii*) with fang-like teeth and a bioluminescent lure to attract unlucky prey. All anglerfish of this and many other related species are female, sparking a long search for the males. They were eventually found, like lost keys, in an obvious place. In this case, he is attached to the bottom side of the female, and lives his life as a brainless, gutless parasite, reduced to little more than testes. See Chapter 10. Photograph by Edith Widder, ORCA.

More than a mile and a half down in the Pacific Ocean, giant tube worms (*Riftia pachyptila*) live at the edge of scalding water that surges from Earth's crust. The worms have no digestive system or mouth but show one of the fastest growth rates of any ocean animal—reaching 6 feet long in less than 2 years. Their food comes from symbiotic bacteria that feed off the high-energy sulfurous water in these hydrothermal vents. See Chapter 4. Photograph by Emory Kristof / National Geographic Creative.

These mating seahorses show one of the most striking role reversals in ocean reproduction—the female here is transferring eggs to the pouch of the male, after which he will brood them and release them as fully formed miniature offspring. Parenting is rare in the sea, where eggs or larvae of fish and invertebrates are often cast out to develop on their own, and male parenting is even rarer. But the whole clan of seahorses has adopted this Father Knows Best approach. See Chapter 10. Photograph by George Grall/National Geographic Creative.

The Pompeii worm, *Alvinella pompejana,* holds the record for the hottest habitat for an animal. The tail end lives at 66° C (150° F) in the chimneys of hydrothermal vents in the deep sea. The head end, an inch or two away, lives at the normal temperature of the deep sea, about 4° C (40° F). How the worm manages to live at both temperatures, and everywhere in between, is a mystery that has sparked a detailed search of its genome for temperature-tolerant proteins. See Chapter 8. © DeepSeaPhotography.com.

A submersible descends into a volcanic vent off Las Gemelas seamount near Costa Rica.
Seamounts like this are home to large fish of great age, some living a century or more.
Photograph by Brian J. Skerry / National Geographic Creative.

This photomosaic shows a whale carcass in an advanced state of decay soon after it was discovered by researchers from the Monterey Bay Aquarium Research Institute (MBARI) nearly 2 miles deep in Monterey Canyon, off the California coast. The large numbers of red worms carpeting its body are osedax worms (*Osedax mucofloris*), the "bone-eating snot-flower." The small pink animals in the foreground are scavenging sea cucumbers. See Chapter 4. © 2002 MBARI.

A healthy coral reef is patrolled by herbivorous fish, like this school of convict tangs on Ofu Island, American Samoa, that chew away algae that could displace corals. See Chapter 11. Photograph by Steve Palumbi.

One of the largest plating corals in American Samoa is also one of the most heat resistant. It lives in a shallow protected backreef that experiences water temperatures much higher than this species, *Acropora cytherea*, can usually survive. Heat-resistant corals also live in the hottest parts of the Red Sea and perhaps elsewhere in the Pacific. They might be the seeds of survival for some corals in the face of ocean warming caused by climate change. See Chapter 11 Photograph by Dan Griffin, GG Films.

A flying fish photographed with its tell-tail zigzag track in the water. Flying fish do not actually fly, but rather glide as the fish attempts to escape predators from below. They gain a propulsion boost by beating the lower fork of their tail along the surface of the water at 50–70 beats per second. The fastest swimming fish can keep up with them, however, and might be lucky enough to snare them as they inevitably come back down. See Chapter 7.

A male sheephead wrasse, *Semicossyphus pulcher* (at the center of the picture), starts out life as a female and changes sex   and color and shape—later in life to control a harem of younger females. Large females turn into males when a local male dies. Photograph by Phillip Colla.

Ocean pout (*Zoarces americanus*), a type of eelpout, live in frigid waters of the Atlantic and have an antifreeze protein in their blood that helps keep ice crystals from forming. The protein is now used in ice cream to benefit storage and consistency even in low-fat products. See Chapter 9. Photograph reproduced courtesy of Animals Animals / SuperStock.

One of the world's largest corals, estimated to be more than 1,000 years old. From Ta'u Island, American Samoa. See Chapter 6. Photograph by Rob Dunbar.

Between the ice and the stars: ice floats atop the Ross Sea, but the water is so cold that ice crystals form on the seabed and reach toward the surface. See Chapter 9. Photograph by John Weller.

of fish otoliths, and mammalian bones are reworked and rebuilt so continuously that bomb carbon does not form a layer in them. So researchers needed to examine this problem through a different lens.

Jeffrey Bada, an ocean chemist at the Scripps Institution of Oceanography, clocked the age of bowhead whales in a completely different manner. His method takes advantage of two peculiar facts about the animals' eyes. First, the core of the lens of the eye is constructed of proteins that are made before birth. Second, once made, the amino acids that make up these proteins slowly change from being 100% left handed, to 50% left handed and 50% right handed. Left and right handedness refer to the exact way that amino acids are formed chemically. All proteins in all cells on Earth are made with 100% left-handed amino acids, but over time, some spontaneously shift from left to right handed. So, if you measure the percentage of right-handed amino acids in the lens of an eye, and you know how fast the amino acids spontaneously shift from left to right handed, you can estimate the age of any mammal.[19] Out of 90 bowhead whales investigated this way, 5 large males were estimated to be more than 100 years old.[20] So, two very different approaches to measuring age give the same result—bowheads can live to be far, far older in the wild than any other mammal known.

Bada and colleagues speculate about why bowheads break the rules of mammal life span. Why do they get so much older than other mammals or even other whales? Their guess is that it boils down to the extreme cold of the Arctic Ocean and why it is so beneficial to be big. Mammals in cold water maintain their body temperature through metabolism and insulation—metabolism comes from body tissues, and heat is lost through the skin surface area. Now, the bigger you are, the more body mass you have to generate heat, and the less surface area you have per body weight. This boils down to saying that the bigger you are, the easier it is to stay warm, and the easier it is to grow bigger still. As a result, bowheads pile as much energy as possible into growth, delaying their first attempts at reproduction so long that they have one of the oldest ages at first pregnancy (26 years) of any mammal. Growing large takes a long time when your main habitat, the Artic, is frozen and foodless for much of the year. For a bowhead whale, the first few decades of life are just a prelude to the vast span ahead.

When whalers stripped the Arctic of bowhead whales, they created a harsh new reality in which living long was not possible for bowheads. During whaling's "golden age," the bowhead whale was one of the most aggressively hunted species. Relatively slow and stocked with blubber and oil, the Arctic giant was hunted nearly to extinction. Europeans lit whole cities by burning their blubber and used their baleen in the umbrella ribs and corset stays employed in elegant Paris fashions.

The populations of bowhead whales began to increase after large-scale hunting of them was banned.[21] But recovery has been slow, 3–5% a year in the fastest growing populations in the western Arctic Ocean.[22] Bowhead offspring gestate for more than a year and are nearly always single births—themselves enormous trials. Pregnancy and milk production for a calf so depletes blubber reserves that a mother cannot produce a calf more than once every 3 or 4 years.[23] With the new age estimates, a female might have thirty calves over 80 years of reproduction—a much greater individual contribution to the population from each female than once thought. And this means that every female is much more important to future population growth—because she has a much larger reproductive potential.

The long lifespan of bowhead whales is most amazing because of what it tells us about the normal lives of whales—before we humans were such a threat to them. In nature, the rate of biological decay with age—called senescence—varies a great deal among animals: a royal albatross can live 58 years, but the similarly sized Canadian goose lives only half this long.[24] But no matter what the possible lifespan, in the wild, few animals show signs of advanced aging—they rarely live long enough to gather the aches and pains that we humans get with our technologically extended lifespans. Instead, most animals are killed by predators, parasites, disease, or extreme weather before they ever grow old enough to show signs of aging. And the animals that have low longevity, even in captivity, tend to be those with high mortality rates in their natural environments.[25] The great longevity of bowheads is a signal that before whaling, bowheads probably had extremely low adult death rates: so low that the aging rate of this huge mammal was adjusted by evolution to slow down to the barest ticking of the biological clock.

Sea turtles

Imagine yourself on the Kona Coast of Hawaii along a shoreline called Puako. The Sun descends into the water at the horizon, and the summit of Haleakala stands above orange-and-purple clouds across the Maui channel. Waves ripple at the black lava shores, and farther out to sea, the water turns from an iridescent sandy-bottomed teal to dark pelagic blue. It is winter in Hawaii, and a humpback whale passes by—then two, then three. Just beyond the breakers, the whales

Green sea turtle *Chelonia mydas* portrait, photographed near Marsa Alam, Egypt. Photograph by Alexander Vasenin.

head south. You snorkel out from shore, snaking through volcanic boulders and breaking waves. Once out to the reef, diving under the rhythmic chop of wind-blown water, green sea turtles—*Chelonia mydas*, or *honu*, in Hawaiian—glide and dive, eyeing you awkward humans with pity rather than fear. One turtle, its shell darker than the rest but streaked with white scars like a painter's dropcloth, lies wedged in the crevice beneath a head of coral. It scrapes algae from the substrate, and you watch its throat work in rippling, snakelike contortions.

That evening *honu* pull themselves up on the rocks in front of your little beach house and bask in what warmth the sunset still yields. You can smell the fish on the grill—fresh opah, caught this morning just offshore and bought this afternoon from a roadside stand. Dinner will be ready soon, and you hold a flute of chilled wine. Drops of condensation trace veins down the glass.

This is a lovely vacation in a tropical paradise, surrounded by beauty and exotic life. But here is what you have not considered at any point in this perfect day: every animal you remember was probably older than you. The humpback could be as young as 20 and as old as 90. The feeding green turtle, with its battered shell, likely ran a desperate gauntlet of ravenous seabirds while your father was learning to shave. The coral under which it wedged itself took hold after the Hawaiian tsunami of 1946 devastated the reef, and the formation on which it grew was constructed centuries earlier from the skeletons of prior residents. And the opah fish you're about to enjoy, with sides of green salad and sticky white rice, likely carried the radioactive legacy of Operation Castle Bravo in its bones. You live in a world embedded in time, and you are not the wise elder. They are.

Animals are tuned by evolution to live long when there is a good payoff to longevity. For a weed in your lawn, the payoff is fast reproduction: it will not live through the winter. But by contrast, sometimes an investment in long life is a better evolutionary bet. Sea turtles have rolled the dice for life, and come up with a paradoxical strategy that has worked well for them for hundreds of millions of years, pairing long, safe lives with a frantically risky reproductive strategy.[26]

Sea turtles adults grow to great size and greater age: the largest species can reach 2,000 pounds.[27] They rely on the classic turtle adaptation: thick shells of layered bony plates, decorated with mottled whorls of green, brown, and black. The oldest carapaces are often studded with gaping barnacles, fringed with delicate fronds of algae, or abraded with jagged white scrapes. But unlike their terrestrial cousins, sea turtles can't pull their heads or fins inside. They wear their shells like bulky Kevlar vests.

Retreating into an impregnable fortress of bone makes sense in terrestrial or shallow freshwater environments, but in the vastness of the sea it is folly. A hunkered turtle has all the swimming capacity of a cinderblock—which is to say, it sinks. A sea turtle's carapace thus becomes a passive defense. Unexpected speed, agility, and a thick shell protect it from all but the most savage predators—sharks and killer whales, lethal but rare. Though humans have hunted sea turtles to near extinction in many parts of the world, their natural adult mortality rate (without human help) is extremely low.[28] Once they reach maturity, sea turtles should have an easy path.

Adult sea turtles live a long time, but their ages are largely a secret. Turtle biologists have finally confirmed that growth lines in some turtle bones are laid down annually, based on dye-injection of some captured turtles and their recapture years later.[29] Growth is fast at first: a juvenile Hawaiian *honu* grows 1–2 inches a year. But a mature turtle in its 30s grows much, much more slowly, making it difficult to assign age just by measuring the shell. Without bomb carbon as a guide, or hidden ancient spearheads, turtles tend to keep their exact ages to themselves.

Sea turtles may live long like whales, but they breed far more abundantly. After mating, females journey hundreds or even thousands of miles to lay their eggs on specific beaches—often the beaches on which they themselves hatched. Under cover of dark, a mother will haul herself out of the water. She starts digging in the sand with her flippers, laboring away for hours before depositing a huge clutch of eggs into the hole. Even young females will lay scores of eggs; larger mothers of certain species produce hundreds.[30] Afterward, the mother buries her eggs and packs down the sand to further hide them. Having spent an exhausting night on the beach, the mother eases back into the sea—and abandons her young forever. She will not return to this beach for years, until her next clutch is ready.[31]

The young gestate under the sand, sustained by the Sun's heat, and when the time comes, they all hatch simultaneously. Beaks tear through leathery eggshells, and tiny flippers start digging at the sand overhead. Hatchlings boil out of the burrow by the dozen and the score and the hundred, and like an infantry platoon, they make a dash for the water. This is no sugar-coated scene from a Miley Cyrus movie; it is a macabre hybrid of the Battle of Dunkirk and the Katyn Massacre. The baby turtles run a gauntlet of crabs, land scavengers, and every hungry seabird for miles around. So they run, and by the dozen and the score and the hundred they are mercilessly devoured. Only a handful will even make it to the water, where additional deadly predators await.[32] The lucky few swim hard for seagrass beds or sargassum forests, where leafy foliage will shelter their growth.[33] Those that survive the trial of youth can expect to live 50 years.

Turtles have achieved remarkable success as a group by getting death out of the way early. Soft eggs laid in defenseless clutches, helpless hatchlings massacred on dark beaches—death is built into the early stages of turtles' lives and is conspicuously absent from the later ones. They mature late and reproduce rarely. Biological aging is nearly suspended for these creatures: the cells and organs of century-old animals are indistinguishable from young adults.[34] These tactics seem to work: sea turtles have swum the seas for 200 million years, and they are among the most enduring and widespread species in the world's oceans.

Black corals: The oldest known animals

Found in deep-sea beds 1,000 feet down, black corals bide their time in quiet darkness. And there is an awful lot of time to bide, for these corals can live for thousands of years. Typical shallow water coral colonies are highly productive and fueled by sunlight; black corals slow their metabolisms to a crawl, experiencing time in a kind of fugue state, with centuries clicking by like years to a human.

These black coral colonies build themselves at a glacial pace—only a hair's-width per year. The result is the painstaking engineering of the ocean's most delicate yet astonishingly enduring creatures. Black corals assemble calcium carbonate crystals into limestone snowflakes, and then twist them together into impossibly delicate branches, tendrils and spines. Called "black" because of their coal-dark skeletons, the polyps themselves blossom off the bone in bright colors: oranges and yellows adorning fine black needles. A few have white polyps, shining in the abyssal dark like evergreen trees after a midnight snowfall. All are fragile, like blown glass sculptures, and are found only where the water is cold and calm.[35] If subjected to strong currents or even the smallest wave, a black coral would be smashed to bits.

In 2009, researchers explored a forest of large colonies of the black coral, *Leiopathes glaberrima*, living in deep water off the coast of Oahu, Hawaii.[36] *Leiopathes* looks like a gangly explosion of orange wire, 3–6 feet high, with bright orange polyp-flowers spread across comb-like branches that sprout in chaotic tangles from tough black stems.[37] The oldest specimens elongate branches at about 1/64 of an inch a year, about the width of four hairs. Their branches thicken by less than 2/10,000 of an inch per year (1/20 of a hair's width). Isotope aging of the skeletons showed that these simple animals had been living and growing since the pyramids of Egypt were built, roughly 4,600 years ago.[38]

These animals take the lesson of the turtle to an even greater extreme. Their larval and juvenile phases live dangerous lives. But once they grow to a minimum size, there is little to threaten them in the deep, constant currents of the abyssal sea. If they grow to be a century old, then they are very likely to live for a millennium, and beyond. And as they grow, they are able to reproduce more and more successfully, releasing more and more larvae. The payoff for slow steady growth is slow steady parenthood, ticking the centuries away . . . and maybe over 4,000 years making a successful coral offspring.

Finding these deep elders is finding a living time machine. They contain memories of the past—not in stories they can tell their grandchildren at dinner—but in cell layers that record and store data about the surrounding environment. Their layered branches open the door to ancient environmental records stored in layers of organic limestone. Any radiation in the surrounding water, the chemical makeup and salinity of the ocean, even year-to-year fluctuations in the climate—everything gets baked in.[39]

Immortal jellyfish

Earth's oceans are so vast and diverse that something in the ocean will prove an exception to almost every biological rule. This chapter has been based on a simple rule: everything that lives must eventually die. Enter the exception: *Turritopsis nutricula,* the immortal jellyfish.

Turritopsis are tiny translucent jellies, just a quarter of an inch across at the base of their bell-shaped bodies, wrapped in sinuous tentacles that sting prey and deliver food. The jellies' lives are pretty boring: eat, reproduce, repeat. Life is brutal for small ocean invertebrates, and predation or environmental stress quickly claims them. *Turritopsis* are dubbed immortal not because they don't die, but because they needn't. They possess the ability—unique among the more familiar animals on the planet—to age in reverse.[40]

Turritopsis start life as a collection of anemone-like polyps that grow across the seabed, multiplying into a lace-like colony with tiny towers of tentacles extending out to catch passing food. The polyps also produce buds that break away and become a medusa, a swimming jellyfish. In turn, the jellies develop gonads to create the next generation of polyps, and then die. All of which is perfectly normal for this kind of jelly.

Turritopsis throws this cycle into reverse. Following injury or the introduction of a new environmental stressor, a swimming *Turritopsis* medusa converts its bodies' specialized cells, even its gonads, back into their original larval forms.[41] By rapidly breaking down its own body, the jelly reverts from an adult to a larva. From there, it can grow up all over again—with a new body, better adapted to its surroundings.[42] *Turritopsis* resemble nothing in fiction so much as the hero of the BBC television classic *Doctor Who.* When close to death, his body "regenerates" with a different appearance and personality. Like the Doctor, these little jellyfish can restart their life cycles following a trauma that fails to kill them outright.[43]

This process is called transdifferentiation, and it allows *Turritopsis* to sidestep senescence. An animal capable of reverting to a juvenile state need never grow old, and thus these animals are technically immortal. We've learned a great deal about the molecular process of transdifferentiation, but little is known about the actual behavior of this species in the wild. Transdifferentiation has never been observed outside a laboratory even in *Turritopsis*. Tiny and translucent, the immortal medusas are nearly invisible in the water column. No individual has been tracked long enough to determine whether *Turritopsis* really use this process to avoid the consequences of growing old.

Village of elders

In American Samoa, a tiny set of eastern islands is the first to greet the rising Sun. The Manu'a islands of Ta'u, Olosenga, and Ofu have been home to human communities for more than 2,000 years. Living long and well is commonplace on Ofu: palm trees, sharks, turtles, and the omnipresent fruit bats live their lives in the wind and the sea and fill the islands from the sky to the reef. But on these tiny island outposts in the middle of the Pacific, nothing is as old as the culture . . . and the corals.

A common kind of coral is a golden mound called *Porites,* a great dome of tiny polyps cemented to the bottom in crystal clear water. They rise from the reef like mounds of bread dough, bulging from golf ball to basketball size, and then growing to the size of a car, and then to the size of a house. Each inch of coral is the growth of years; each inch is proof that the reef still thrives. Offshore of Ta'u lives the mother of Samoan corals, a mound that rises from the reef toward the surface and is 30 feet tall and 40 feet wide. It is heavier than the largest blue whale. It is older than any other living thing in Samoa.

It is not alone. Though it is biggest and oldest, it has company around these islands in other massive corals that populate the reefs and lagoons like massive vehicles scattered across a living parking lot. The easiest to find are on Ofu Island, in shallow back reef pools no more than 10 feet deep. There, in a village of elders sits a rank of massive corals. One huge colony after another lives and grows its slow inches from decades to millennia. Some colonies are so large they have exhausted the depth of the lagoon, long ago growing tall enough to hit the surface. No longer able to grow higher, they continue to grow wide, and form squat barrels of coral 30-40 feet wide.

The lagoon that houses the village of elders is a deep well of time, remarkable and precious for its constancy. Each year each tiny coral polyp encrusting every square inch of these colonies feeds, grows, and makes another few layers of calcium carbonate skeleton. But each year across the centuries, each day across the years, one environmental disaster could spell the doom of every living elder. The magic of this village, and the record of the elders living in it, is that the lagoon of Ofu has supported vibrant coral life for every single day of the past millennium. Coral have the intrinsic ability to live long, growing into the future like no other living thing. But the reef and lagoon must let them, and must be a constant source of living support.

CHAPTER 7   THE FASTEST SPRINTS AND LONGEST JOURNEYS

*The drag of water makes both speed and distance a challenge
in the marathon of the ocean.*

Marine creatures are attuned to the drag of water in ways that we simply can't be. We're land creatures, striding from place to place as though in a vacuum, taking little note of the feeble atmosphere swaddling us. Water presents a greater obstacle to movement. The density, the weight, and most importantly, the way water clutches strongly at everything moving in it makes it a permanent, physical impediment.

The drag of water pulls at everything that moves in the ocean. Drag, and the stamina to withstand it, are important in the fastest sprints when fish muscle their way up to 40 miles per hour. Or when squid pump seawater through their bodies to become natural jet engines that fly. But drag and stamina are also a cruel tax during the longest treks of any species on our planet, journeys of tens of thousands of miles. Whales swim with a plateau of efficiency that no human machine can match. Even the albatross, soaring the greatest weather systems on Earth, sips strength from the waves to power their epic migrations.

The speed of a herring

A competitive swimmer, leaping from the block, hits the water with a shock. His heart jumps. His skin crawls. He keeps his head down, tucked into the prow between his arms—if it's raised even a little, the water will rip the goggles off his face. At the moment of impact, his hands rip a corridor through the pool. As he finally breaches the surface to breathe, his hands split and arms twist into a new configuration. The movements are lightning fast, as powerful as young muscles can make them, and expertly angled. He labors to deny any grasp to the reaching water. Every movement trains the fluid around him; his muscles flow in a partnership with water. The fastest swimming Olympian burns his body to move at 5 miles per hour—slightly faster than a herring.[1]

The comparison is not fair. The human athlete has spent a life tuning a terrestrial body to perform in an alien, liquid environment. The herring is at home in the sea, and its evolutionary legacy is a set of adaptations that help solve the enduring problems of moving through water.

Fastest under sail

A sleek fishing boat rolls over an ocean swell. Twenty-five feet long, snow-white fiberglass shining in the Caribbean Sun, it's a craft for tourism and commerce. But it is also a technical marvel. Powerful engines, advanced reels, fishing lines crafted to high-tech perfection—modern technology has expanded the pursuit of the ocean's fastest fish, matching muscles, power, and speed with materials, power, and speed.

A fishing guide scans the sea from behind black mirrored lenses. As the vessel crests yet another swell and the prow plummets, he lets out a whoop. He points a yellow-gloved finger to a ripple in the water, a hundred yards out. Straining their eyes, the passengers can make out only a churning strip of bubbles. And then the fish jumps—as if on cue, to aid novice eyes. Silver and black, with a spear-like bill and gleaming scales stretched across slabs of rippling muscle, a sailfish soars into the sunlight. For an instant it hangs in the air above a galaxy of silver droplets. With a white explosion, it's gone below, streaking just beneath the surface, its portrait split into a hundred Cubist slivers.

Sailfish, of the genus *Istiophorus*, are arguably among the ocean's greatest natural athletes. They combine a streamlined muscular body and a tapering long menacing beak. All the members of the family of billfish—counting sailfish, marlins, and swordfish in their number—bear a strong resemblance to one another. They lead similar lives: large, solitary predators who stalk the continental shelves of the world's oceans for smaller fish. Billfish are very fast, but the sailfish has adapted speed beyond any of its cousins. They are said to streak at 80 miles per hour during jumps and to hunt their overwhelmed prey at a consistent 30 miles per hour.[2] The combination of fins and muscle—geometry and physics—gifts them unparalleled efficiency.[3]

Hatching from eggs thrown into the open ocean, sailfish begin life with an astonishing growth spurt. In their first 6 months of life, they'll grow from specks to more than 4 feet long.[4] Even when young, sailfish stand out for their distinctive "sail": an enormous, fan-like dorsal fin that raises and lowers like a fan. It's collapsed as the fish screams through the blue, but can spring to attention instantly.[5]

Careful analysis suggests that, just as a human sprinter does not run everywhere at full speed, billfish save their speed for special circumstances.[6] Even feeding can be more orderly than once thought. Gilbert Voss, a fish biologist in Miami, described a famous event in Florida in 1940, when sailfish were seen to feed for the first time. There were six to thirty sailfish in each group, circling small herring-like prey to force them into tightly packed schools, scaring them into compliance by raising and flashing their huge sails. Once the ball of fish was nicely formed, one after another, the sailfish would strafe through the schools, stunning fish with up-and-down and side-to-side swipes of their bills, and snapping up stunned prey.

Monitoring fish speed is notoriously difficult. Ocean life writer Richard Ellis relates that the oft-cited quotation for sailfish speed (68 miles per hour) comes from the Long Key Fishing Club in Florida, where fishermen conducted experiments with hooked sailfish and a stopwatch.[7] Precise measurements of fast fish have been made a few times and have clocked some sprinters at the high speeds we would expect. Wahoo, a torpedo-shaped relative of the tunas and billfish, has a top speed of 48 miles per hour, similar to that of a yellowfin tuna (47 miles per hour).[8] Some fish are said to go faster, and probably can. But moving that fast requires more than just muscle.

Fast food

All that velocity also demands a body able to *eat* at high speed. Hunting at billfish speed is tricky, like driving at 40 miles an hour on a busy street while trying to snatch a coffee mug off the asphalt. Roaring through the open water, billfish plow through schools of prey with quick twisting movements, flashes of the eyes, and calculated swipes from sail and bill. Chewing is a waste of time; swordfish suck down their dazed prey with gulps like fingersnaps and don't even have teeth as adults.[9]

All billfish and their cousins the tuna are cold-blooded animals, and they live in cool ocean waters that teem with food. But they nevertheless keep their muscles warm through strong exertion. Tunas even have a heat trap in their blood stream that limits how much warmth they lose through their gills.[10] To generate heat besides that provided by muscles, these fish have evolved specialized tissues that function as heating units. They're muscles with no ability to contract: dark brown flesh that converts calories directly into heat rather than motion. You may see them in a tuna as the dark brown flesh on either side of the spinal cord. But even if the core of the fish's body is warm, other parts, especially close to the water, like the muscles of the eyes and the reflex nerves, stay cold and function slowly. In billfish, where eye and nerve reflexes are critical, the brown muscle heaters are also found where they're most needed: right next to the eyes and braincase.[11]

Maintained at warmer temperatures, often more than 7° F (4° C) warmer than the surrounding water, the eyes in particular can operate at race-car speed and precision—the kind of suspended "bullet time" typically available only to Hollywood action stars. The retina of a swordfish can process information fast enough to detect the quick flash of prey fish during a high-speed pass. But if you cool down the retina enough to match the temperature of the ocean, its scanning speed drops to the point where a fast flash of prey is invisible.[12] Warm retinas also see better at low light levels, meaning that eye heaters give swordfish an advantage 1,000 feet deep, where they often forage. So these predators enjoy razor-sharp vision and lightning-quick reflexes as they drive through schools of darting fish in a cold sea.

Flying fish

Skipping across the sea behind a Fijian barrier reef in a high-powered skiff, you unknowingly run down a school of 10-inch-long fish. Instead of scattering, they rocket ahead, erupting from the water in a torrent of turquoise. Fins held akimbo like wings, their tails thrash in furious sine waves—propelling the small creatures just as fast as your boat. With an air of bored disdain, they bank hard left in stunt-pilot unison and head for the horizon.

Powered flight evolved three separate times on planet Earth: in birds, insects, and mammals. In each case, evolution devised a different method for translating muscle movement into aerodynamic lift: wings with feathers, wings of exoskeleton, and wings of skin. While birds, insects, and mammals evolved flight in their own ways, a fourth model labored hidden under the waves. It was a one-off, an aerial exotic, an evolutionary miracle that has been given the ocean's most appropriate name: flying fish.

More than fifty species populate the tropics.[13] Their torpedo-shaped bodies are well muscled all along the fuselage. Their strength provides speed, but the density of water makes high speed an extravagantly expensive metabolic proposition. The power needed to move through water or air increases as the square of the velocity.[14] But air is much less dense than water, and the drag air creates against fast travel is far, far less. So, fast travel is expensive in water, but cheaper in air. In response to this simple physical fact, the flying fish, family Exocoetidae, evolved their fins into technological marvels of flight.

Those fish seen from your skiff took flight for a reason. Imagine them just a few moments before, cruising in a loose echelon just below the water's surface—gorgeous animals proudly marked with blues and purples and yellows. Their pectoral (shoulder) fins are the most eye-catching feature. Elongated into wings, stretching to either side, their delicate spines and translucent planes resemble insect wings: butterflies birthed in the sea. The school swims in formation, feeding on plankton and tiny fish.[15] Suddenly a mahi mahi appears from the dark Tartarus below, 20 pounds of needle teeth and muscled fury on an intercept path. The formation's left outrider—a little cerulean female—is the closest to danger and the first to turn. Her companions follow, veteran wingmen instantly sensing the shift.

She pushes hard, but the mahi mahi is utterly committed. It's bigger, stronger, and—over a 100-yard sprint—faster than our heroine. She's got nowhere to go but up—and accelerates to 20 miles per hour,[16] angling up toward the surface. Death churns inches behind. Reaching her world's silver ceiling, she leaps and ignites a hidden engine—a tail with an elongated lower fork. The lower lobe churns water at 50–70 beats a second. The ultimate effect is that of a rocket.

Our heroine spreads wide her pectoral wings. A second set of smaller fins near her tail splay out too, making her like a biplane. As pectoral fins provide lift, the tail delivers power. With one final push from the long lower lobe of her tail, the only part of her still in the water, the sea abruptly falls away. The air cushions her now, hot and suffocating but simultaneously life-saving. Her wingmates form in on either side. The squadron hums low across the surface at double their waterborne speed, turning abruptly, to leave the predator behind.

But the mahi mahi isn't ready to quit. It is capable of bursts of 40 miles per hour, following along below the surface.[17] And although it seems as if the flying fish has escaped entirely, there remains an enduring problem. The problem is that flying fish do not fly—they glide. Their fins are not beating wings that provide powered flight; they are only gliders that merely prolong flight. And once out of the water, the flying fish—and the mahi mahi—are faced with the inevitable fact that the gliding flying fish will come back down.

Flying fish pursued by a dolphin fish. Published in 1889 in *Popular Science Monthly,* volume 35. Author unknown.

Each flight is only seconds long. Cruel gravity pulls her toward the water—and beneath is the rainbow form of the mahi mahi matching speeds. When she starts to lose altitude, she dips her tail and tags the sea for another burst of power. Up to a dozen times she might graze the surface and glide again.[18] The death race flashes across the sea, covering the 50-meter length of an Olympic sized pool in seconds. The mahi mahi tracks the flyer, keeping up with the slow curves of its evasive maneuvers. It's eaten many flying fish, and every new dip of our heroine's tail offers a fresh attack window.[19] But a missed strike will let the flyer escape, and so the predator holds fast. Both fish are at their physical peaks racing toward the horizon.

Only an abrupt splash in the distance tells you that one or the other has won.

Jumping dolphins and bumpy fins

Nothing in the ocean has quite the exuberance of a pod of dolphins on the move. Powering just under the surface, they slip through the water with barely a ripple, then leap into the air in a graceful arc, almost as if to give us poor landlubbers a good look at their beauty. But is it just impetuous joy that drive the leaps of a running dolphin? It turns out that it may be good common sense.

Rapid swimming is expensive, increasing in cost as the square of the swimming speed. By swimming and leaping, perhaps a dolphin is taking advantage of the lower drag in air compared to water—maybe the occasional leap reduces drag to save calories. But leaping involves a lot of energy too, so for leaping to be worthwhile, the cost of the leap must be lower than the energy saved by arcing through the air. A careful analysis confirms this, showing that leaping is only worth the cost above a critical speed threshold, the "cross-over" speed. Dolphins swim smoothly just below the surface, up to about 10 miles per hour. Above this speed, leaping becomes more efficient than swimming, so dolphins jump.[20] It's more than fun, it's simple economics. Size matters in this equation: the bigger you are, the more costly is a big jump—and the higher the cross-over speed should be.

The "great" whales are too big to travel this way. It costs a great deal to throw 30 tons of whale out of the water, and cross-over speeds would be in the range of 30 miles per hour. They rarely reach this speed while migrating. But some whales leap anyway. The question is, why?

Humpbacks (*Megaptera novaeangliae*) are medium-sized whales, up to about 60 feet long. They're common and populous, have rebounded substantially from whaling, and are beloved by tourists for their legendary breaching behavior.[21] With mighty strokes of their flukes, the giants hurl their upper bodies out of the water. Backs arched, they smash down in an explosion of glittering white spray. Fin whales, significantly larger, have been known to perform the same herculean leaps,[22] but most of the biggest whales do not. What the humpback really boasts over its cousins is agility. Breaching requires contortion to a degree that freight-train blue whales never attempt. And a humpback's pectoral fins are the key to its sinuous movements.

Proportionally, they're the longest of any cetacean—18-foot fins aren't uncommon on animals 65 feet long.[23] Delicate like angel wings, streaked with white, are studded along their leading edges with knobbly protuberances often mistaken for barnacles. These are actually hair follicles, modified and swollen to rubbery fist-sized bumps called tubercles. The larger a fin is, the more water runs along its leading edge and the more drag it incurs. Tubercles, unique to humpback whales, break up the current to force water across the fin instead. The end result is a vastly improved hydrodynamic profile and increased underwater lift.[24]

Human engineers eventually took notice, admiring whale fins and learning the lessons they taught. A Canadian engineering firm recently designed wind turbines modeled after whale fins, with metal ridges cut like tubercles

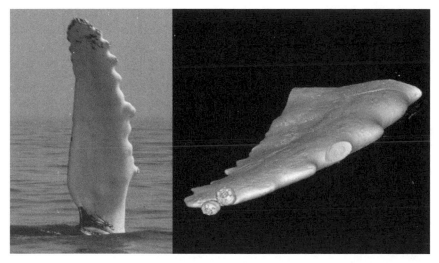

(Left) View of humpback whale (*Megaptera novaeangliae*) pectoral flipper showing leading edge tubercles. Left image courtesy of W. W. Rossiter. (Right) Three-dimensional reconstruction of flipper tip from CT scans. Both images from Fish, F. E., L. E. Howle, and M. M. Murray. 2008. "Hydrodynamic flow control in marine mammals." *Integrative and Comparative Biology* 48(6):788–800, figure 5. Used by permission of Oxford University Press.

into the leading edges of the blade.[25] They move air with absurd efficiency for such a slightly improved design: 32% less drag and 8% increased lift compared to a smooth model.[26] A whale-fin cooling fan spinning at 16 feet per second moves as much air as a typical fan working 25% harder, a thrilling improvement for engineers used to small marginal gains.

Jet propulsion

The speedsters we have described so far all have bones. They carry their bodies' skeletal structure inside, like fish and marine mammals. Other fast actors we'll get to next rely on armored exoskeletons. But in both cases, muscles are anchored to hard components. Those hard components transmit power to bones or shells, which move water and generate thrust.

But some of the most numerous mobile creatures get by without skeletons. Squid are the finest example: without a single bone in their bodies, they slash through the seas powered by nature's own jet engine.[27]

Squid are *cephalopods*, "head-foots," members of a large class including cuttlefish and octopuses. Large, prominent eyes take in the world with a cold alien intelligence. Eight arms and two longer appendages (properly called tentacles to differentiate them from arms) dangle below the eyes. They typically surround a cephalopod beak. The end result is a giant eye-defined "head" married to a confused, tentacled "foot." Behind this head-foot, a squid's mantle is a long muscular tube containing most of the organs, tapering into a fleshy conical tip. The animals have no teeth or bones, except for some cuttlefish that have the remnants of a coiled shell buried within them as a 'cuttlebone'.

Squid move by pumping water in and out of their bodies. Propulsion comes from using the water itself, sucked into the mantle and squeezed out through a smaller tube called a siphon in a series of strong pulses. By finely manipulating their siphons, squid maintain precise control the water stream: volume, intensity, and direction.[28] All cephalopods carry siphons, even the lumbering octopus, but squid get the most mileage from them.

Water is heavy, so you'd expect slow acceleration from a squid. Not so: powerful rings of muscle surround the mantle, squeezing a huge amount of water through the siphon and creating large accelerations. They've also got a secret weapon for emergencies: a lightning-fast escape mechanism. It's similar to a lobster's caridoid reflex: a highly specialized nerve structure called a giant axon,[29] a single supersized nerve fiber thicker than a human hair running down the mantle. Its unusual size facilitates the extremely rapid transit of nerve signals to all the muscles of the mantle, so that the maximum amount of water will shoot from the mantle and jet the squid away. The giant axon is a way to translate a simple imperative ("Run!") into a complex escape response.

The waterborne jet propulsion of squid has been well known for centuries. Since the 1800s, observers have described squid gliding *above* the water. It was a marine legend, but a recent survey recorded a group of squid off the coast of Brazil doing more than gliding. A fan of 6-inch silver missiles burst from the water and accelerated away, trailing streams of high-pressure water.[30] A second report concerning Japanese squid shows the same capability in ocean-going animals best known as a fishery.[31] The aerodynamic details are not yet completely understood, but jet propulsion allows a flying squid to do something flying fish cannot: accelerate in the air. The Brazilian species *Sthenoteuthis pteropus* accelerates at more than two g's. However, fuel runs out quickly—limited by how much the mantle can store—and such brief acceleration can't get the animals past about 8 miles per hour while airborne. It's a rare testament to the power and grace of these animals—a phenomenon few people have ever witnessed.

Lobstering

Crustaceans are among the ocean's most common and successful creatures, but they're still so *awkward*. Consider, as David Foster Wallace once did, the lobster. Cursed with legs in a world of fins, he scuttles across the bottom while fish cruise above. Weighed down by interlocking plates of armor, he turns like an old station wagon. If bowled over by a passing wave, he thrashes around on his back, waiting for the next wave to save him.

That heavy carapace has advantages: large adult lobsters have few natural predators. But as juveniles, they're devoured by a host of threats from octopuses to their own grown relatives.[32] Smaller crustaceans like shrimp face predation all their lives. In response, many crustaceans evolved a stunning adaptation: an escape reflex so powerful that it single-handedly catapults these scuttling tanks into the realm of elite speedsters.

Our chitinous subject is under attack, or thinks he is—a diver approaches, holding out her camera for a great shot. At this provocation, he suddenly contracts, pulling his tail down and around to his underside in a series of rapid-fire spasms. Extending back, curling down from the thorax, a crustacean's abdomen is built of heavy muscles.[33] The escape motion is like your outstretched hand, palm up, contracting into a tight grip. It launches the animal backward in an eyeblink: a medium-sized lobster can accelerate at 330 feet per second squared,[34] ten times the acceleration due to gravity, seeming to teleport the animal 5 feet back in an eyeblink. A Bugatti Veyron Super Sport, among the world's fastest production cars, manages an acceleration of only about 40 feet per second squared.[35] Like the *Looney Tunes* roadrunner taking off, our lobster is gone—leaving nothing behind but a swirling cloud of sediment. Six days out of the week, his tail is a slab of deadweight best served with Hollandaise sauce. On the seventh day, it saves his life.

The motion is called the caridoid escape reaction, or more simply, *lobstering* (though it has been most studied in the crayfish).[36] It's supported by delicate infrastructure throughout the nervous system, neurons and nerves that propel the animal's whole body into fast action. Crustaceans carry their tails perpetually cocked for lobstering at a moment's notice. When threatened, they can trigger the caridoid reflex in as little as one hundredth of a second.[37] But can their brains really operate that fast? Lobstering, it turns out, is not a choice in the typical sense. Instead it's the product of a *command neuron*—a system integrating thousands of nerve processes into a single hair trigger.[38]

In the lab, one electrical stimulation of the giant neuron, or just a poke at the animal's abdomen, triggers a joint muscle response of artificial panic.[39]

Once started, it just happens, like the effortless and unwavering precision of a Ray Allen jump shot. The giant nerve fibers provide a very fast response—but it is limited in versatility. Through countless repetitions and eons, a simple desperate tail-flip movement evolved into a permanent panic circuit in crustaceans' wiring. Their brains surrender to a hard-wired routine baked deep in their bodies, sacrificing control for raw speed. Among all the world's animals, no escape reflex is faster.[40]

Quickest on the draw

You're on vacation in the Caribbean, taking a snorkel cruise over a local reef. Bobbing on the stepladder, rubber fins on corrugated steel, you bite down on your snorkel and take the plunge.

It's warm as bathwater. Purples and pinks and blues checker the bottom, crisscrossed with the heavy fish traffic of a healthy reef. You can't hear much through the gurgle of water in your ears, but it's impossible not to notice a persistent clicking sound. It's more than a few clicks; they're beyond counting, like a thousand pebbles thrown into a tumbler. Your first thought: currents churning stones on the bottom. But in fact, this is a biological sound—that of a tiny and fascinating crustacean, the snapping shrimp in the family Alpheidae.

But snapping shrimp do not make a sound the way you might think. They do not bang two parts of the claw together to make their signature snaps. Instead, they manipulate the basic physics of water, and hugely amplify the sounds of their rapid-fire claws.[41] They take advantage of *cavitation*, the tendency of water at ultra-low pressure to vaporize into small bubbles. When the pressure falls, the bubbles form, and when it rises again, the bubbles suddenly collapse to release a great deal of energy in a small space and brief time. Cavitation is a major engineering problem in boat propellers—they spin so fast that they create millions of tiny bubbles that collapse with a small bang. That bang eats into the metal of even the hardest propeller blade, limiting its operational life. Snapping shrimp have invented a biological way to create these cavitation bangs, shaping their shockwaves into weapons fired like an old flintlock pistol.

For this reason, snapping shrimp are often called pistol shrimp. One of their two claws is a delicate prong, but other has swollen to enormous size—as though Bruce Banner's fist decided on its own to enter Hulk mode. This claw carries a modified pincer: one side a notched blocky mass and the other

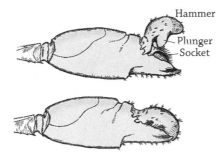

Hammer
Plunger
Socket

Snapping claw of the snapping shrimp *Alpheus californiensis*. (Top) the snapping hammer is held open by a catch mechanism. The moveable hammer, the plunger, and the socket are shown. (Bottom) The snapping mechanism is closed, with the moveable hammer and the plunger fitting precisely into the anvil socket. From Johnson, M. W., F. A. Everest, and R. W. Young. 1947. "The role of snapping shrimp (*Crangon* and *Synalpheus*) in the production of underwater noise in the sea." *Biological Bulletin* 93:122–138.

a hinged finger-like protrusion. Think of that "finger" as a hammer, and its blocky counterpoint as an anvil. The former fits into the latter's notch as though they were shop-machined.[42] The hinge joint has powerful musculature, and the shrimp carries the whole assembly cocked open like a revolver, complete with locking catch. When the shrimp triggers the joint, the hammer slams into the anvil socket.

The descending hammer displaces all the water from the notch reservoir. An intense jet of water shoots forward directed by the notch's lip, moving so fast that cavitation produces an air bubble in its wake.[43] The bubble rockets away, but slows rapidly under the drag of water and grows unstable as the pressure rises. It dies with a pop just an instant after birth, collapsing explosively to release heat and light. Temperatures inside the bubble spike to 8,500° F (4,700° C): more than enough to melt tungsten.[44]

The "pistol" casts a powerful shockwave forward into the water, giving the shrimp a hunting tool. A snapping shrimp lies in its burrow like an undersea highwayman, weapon ready to ambush small prey. Its shot erupts from the tiny aperture faster than any possible reaction and hits the prey with the force of a hammer blow. The flash and moving water are too quick for the human eye, so the target appears stricken backward by an invisible force—as though smote by some vengeful reef god. Having stunned the victim or killed it outright with a single attack, the shrimp scoots out from its burrow and hauls the meal back.[45]

As you might imagine, this unique tool is good for more than violence. The family Alpheidae comprises hundreds of shrimp species, many given to claw wrestling or chesty displays of noise. A rare few have formed hive societies, similar in some ways to social insects like bees and ants. In hives, comrades signal greetings and call out threats with pistol shots, holding their weapons overhead for safety. For such able fighters, the hive-living species tend to avoid conflict. Disputes are resolved with posturing and warning shots, but competitors rarely injure each other.[46]

Long runners: The great migrations of whales

Moving fast in water burns your energy budget at ruinous rates, but another challenge presented by the wide ocean is the possibility of long-distance migrations. From some spots of Antarctica, a gaze due north runs across open ocean all the way to the Arctic Circle. Such broad swaths of habitat are opportunities for very long distance travel—and some ocean species use them. Powering those migrations demands different solutions than the solutions for speed.

Thirty feet deep in the open ocean, the blue ranges in every direction from transparent to a dusky dark that seems to swallow light. Waves march above in a long repeating line, striving for the nearest shore, thousands of miles away. A silence lies throughout the sea like unbroken fog.

Unexpectedly, a shape slips past, giant and dark, with a powerful flat tail and two wide fins splayed out to either side. A second, smaller shape ghosts behind. They rise to the surface synchronously, blow several fishy-smelling notes in quick succession, and sink back to their placid hike. When they are gone, the ocean returns to empty waiting.

Some of the longest migrations on our planet are by swimmers. Blue whales slip from the Southern Ocean near Antarctica to subequatorial seas. Humpbacks announce their annual return to Hawaii, after a long swim from Alaska, with shows of exuberance above the water and majestic operas beneath.[47] Gray whales leave their foraging grounds in the Bering Sea and meander down the coast of California to breeding lagoons in Baja Mexico.[48]

For humans, swimming is wonderful exercise, because water is heavy. Moderate-speed swimming burns as many calories as fast running, rowing, or cycling. As any parent of a competitive swimmer can attest, it takes tremendous food energy to swim. Olympic swimmers may consume upward of 10,000 calories per day.[49] Given these huge costs, how and why do whales manage such long migrations?

Whales tank up before a long swim. Often their feeding areas are in far polar seas, and they spend their summers feasting on the bonanza of food that a short, intense polar summer can produce. But as winter approaches, the water gets colder, and the plankton fade into a fallow offseason. So the feeding grounds are abandoned, and migratory whales turn for the tropics at summer's close.

With full bellies, they still face 5,000 miles of cold water between their food supplies and winter havens. They aren't explosive athletes; if a dolphin is a supercharged cherry red motorcycle, these migratory titans are freight trains rumbling from coast to coast. Like those trains, they take a while to get up to speed. Acceleration isn't nearly as valuable as efficiency, so the giants propel themselves with powerful deliberate strokes. A blue whale routinely cruises at about 1–4 miles per hour, wasting not a scrap of strength or movement.[50] Once a great whale reaches cruising speed, it doesn't take much energy to maintain that speed.[51]

Their size draws the most attention, but migratory whales are marvels of engineering. A blue whale (*Balaenoptera musculus*) is the greatest example, stretching 200 tons of mass out to a surprisingly slender 100-foot length. The tail fluke delivers thrust at 90% efficiency—far higher than the best commercial ship propellers, which churn and grasp desperately for thrust that the whale fluke seems to calmly command.[52] Backing up these traits is a steely endurance, pushing the traveler across entire oceans in just weeks, without feeding.[53]

They arrive on wintering grounds, rest, nurse their young, and sing the melancholy songs that attract mates (whales are big into Morrissey). They choose warmer water, gamboling across the reefs of Tonga, frolicking in Baja's lagoons. Shivering less than they did in the cold polar seas, the energy saved by summering in warm water can recoup migration's costs. New calves likely benefit most from the heat; being small, they lose more heat per unit body mass than their parents do. They also avoid their fiercest predator: Orcas hunt in colder waters and feed on helpless yearling whales when they can. Warm winter waters are a haven against some of these dangers.

The end of winter is a hungry time for a migrating whale—especially a mother who has nursed her young calf, drawing on the food reserves she laid down the summer before. Yet the reverse migration lies ahead for these animals, thousands more miles, this time on an empty stomach. Food at the end of this journey looms as a critical need for even the biggest animals on Earth, and they arrive ravenous in their polar summer seas.

Lately, that food has been harder to find for some whales. In 1999, a large fraction of gray whales (*Eschrichtius robustus*) were observed to be thin, even emaciated. The death rate of calves on migration was huge, and hundreds died. The reason has been traced to the northerly retreat of their food supply: instead of a wide field of tasty crustaceans that gray whales normally find in the Bering Strait, the whales found nothing. Their food supply had responded to warming waters and had retreated to the north. Grays weren't helpless— they followed the trail and found meals at last, at the cost of hundreds more miles and countless calories.[54] Some whales probably didn't find enough food fast enough, and their calves died. Others plowed north into the Chukchi and Beaufort Seas, victims of a warming ocean, looking for food.[55]

Gray whales are regularly seen much farther to the north than they used to be. But one animal stunned the whale-watching world in 2010 by turning up off the coast of Israel in the Mediterranean Sea.[56] No gray whale had ever been seen in the Mediterranean, and the species was hunted out of the Atlantic around the year 1700. For 300 years, this species has been restricted to the Pacific. This one whale may have slipped through the Bering Strait and across the Canadian Arctic ice fields, swimming thousands of extra miles, pushing the normal migration distance far, far longer. Once the wanderer emerged in the North Atlantic with a full belly, it simply followed instinct: head south for the winter and eventually turn east into a nice warm lagoon. This lagoon happened to be the Mediterranean, not a breeding site in Baja.

However it managed the journey, this whale disappeared and has not been seen in either ocean since.

"Albatross!"

Mile for mile, only one animal on the planet can match the great cetaceans. It's a seabird, skimming inches above the water's surface with the vast white wings and inner peace of a seraph: the albatross. Long celebrated by poets as a symbol of natural beauty and by sailors as an augur of death, the bird's image resonates through maritime history. Though hatched on land and bound by memory to strands of beach, albatross are really creatures of the sea. They spend most of their lives in flight, feeding from the water as they cover endless open miles. Living in solitude except for brief stretches during the breeding season, albatross drift across the sea in lazy gyres the size of continents.

Flying in air doesn't have the same energy demands as swimming—the drag of air is far, far less than that of water. But the sheer length of migrations for some seabirds qualifies them as extreme by any metric. The wandering albatross, *Diomedea exulans*, is the largest seabird in existence and boasts the longest wingspan of any living bird—outstripping even the California condor. Those wings, luminous white streaked with sooty black feathers, stretch up to 6 feet each from the slim white body—12 feet in total wingspan. Yet these sleek birds weigh less than 30 pounds; a long hooked beak and dark eye highlights complete their striking appearance.

In the 1977 Disney film *The Rescuers,* the heroes (a pair of animated, anthropomorphized mice) hire an albatross to fly them from New York City to a Louisiana bayou. Their charter, a klutzy aeronaut named Orville, offers great comedic relief with hair-raising takeoffs and slapstick landings. It's neither Orville's fault nor Disney's invention: albatross, collectively, have a rocky relationship with the ground. They don't so much land as crash in slow motion, cutting speed until they can hit the ground without injuring themselves.[57] A fierce headwind helps.[58] Taking to the air is clumsy, too, but the bird has to work much harder. He can't fly until he's reached a decent speed, getting air moving over his huge wings. After some spastic stretching, he starts down the runway with a furious waddling gait and wings a-flap. Some tentative hops, testing the breeze, hoping he gets a well-timed gust. He finally catches some air and keeps it, angling up into the wind. If he avoids an early stall, he's more than airborne: he's home. You don't need a degree in ornithology to appreciate an albatross in flight. Wings stretched and motionless, angled edges perfectly slicing the wind—more gliding than flying. The tranquility of an albatross cruising over a swell at twenty knots is the key to his beauty and his entire way of life.[59]

Flight is easy for birds, but flapping is exhausting. Every stroke of their wings takes energy, so each moment of powered flight needs fuel. Hummingbirds are the prime example of high-energy reliance: a single pause in ever-beating wings would drop them like stones to the ground. Albatross are the polar opposite, accruing 75,000 miles per year with a bare minimum of effort.[60]

Such long hauling wouldn't be possible for a flapping albatross. She'd collapse from exhaustion or starve to death. So she puts those enormous wings to use as a static airfoil, like a hang glider, stretching them out to full extension once she's in the air and keeping them fixed. Most birds' wings would tire after a few minutes; the albatross can endure for days. Special tendons in her shoulders lock them in place, needing no energy whatsoever to hold her wings outstretched.[61] Following large weather patterns over the oceans, the

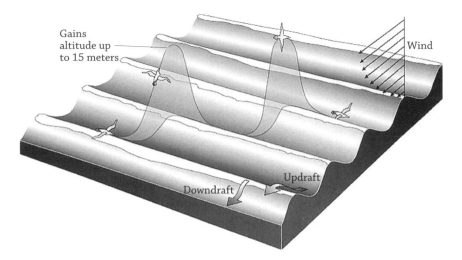

Gains altitude up to 15 meters

Wind

Updraft

Downdraft

How albatross soar. The wind and the waves are moving from upper right to lower left. A bit of an updraft acts on the upwind side of each wave, and the albatross can catch it and be lofted up to 15 meters. They turn to the right and glide down, moving upwind and then along the trough of the wave until they again bank into the updraft. Reprinted from *Progress in Oceanography* 88, P. L. Richardson, "How do albatrosses fly around the world without flapping their wings?" 46–58, © 2011, with permission from Elsevier.

albatross rides for thousands of miles on a cushion of air. But for the slightest twitches of her tail and subtle turns of her abdomen, she is completely inert. Her heart rate during flight is no higher than at rest.[62]

Gliding, even at its most efficient, requires some kind of extra thrust. Albatross find it in the same winds that carried sailing ships between continents, using tactics identical to those of experienced glider pilots.[63] If in need of a boost, the birds dip their wings and lose altitude while rapidly gaining speed. After dipping into the valley between two swells, they turn 90° and bank upward. They turn into the wind, rushing up the peak of the second wave, losing a bit of speed but rocketing up in the air to continue their lonely odysseys.[64] Large, broad wavefronts in the open sea can serve as glider mass transit, with several albatross hugging the front of the swell. They're surfing—not on the water, but on the thin current of air pushed ahead by the wave. By using these tricks in favorable spots, albatross are able to cover more than 500 miles each day.

When the wind dies, unlucky albatross must either burn energy flapping or sit in the drink until their fortunes to change. Physical oceanographer Phillip Richardson lays out the rules albatross must follow "No wind, no waves, no soaring."[65] Becalmings are common in the world's warmest climates, and for that reason, albatross are seldom found in the deep tropics. They're particularly fond of the planet's southern reaches, underburdened with land mass and home to vast empty tracts of cold sea and variable winds. Wandering albatross live long lives—up to 70 years—and may forage over 9,000 miles during a single breeding season.[66] They don't match the awe-inspiring size of the great whales, but the sight of an albatross hanging suspended by the invisible threads of the wind should still take your breath away.

That made the breeze

Sailors long believed the albatross, soaring forever and making its own way through the world, actually created the wind that propelled them all across the vasty deep. In Samuel Taylor Coleridge's immortal poem, the eponymous Ancient Mariner disastrously kills one of the white birds—a beautiful thing with ghosts in its eyes and a cold wind at its back:[67]

> *And I had done a hellish thing,*
> *And it would work 'em woe:*
> *For all averred, I had killed the bird*
> *That made the breeze to blow.*
> *Ah wretch! said they, the bird to slay,*
> *That made the breeze to blow!*

CHAPTER 8 THE HOTTEST

*The oddest thing about warm ocean life is how many species
live near the top of their heat tolerance.*

Deep and hot

Thailand's Andaman Sea heats like a skillet. There's no stove below—rather, sunlight glares mercilessly above, broken up by eerie rock columns into wavering oblong shadows. Every crystalline morsel of water inhales the streams of solar energy, but a whole ocean is hard to heat. And once the surface temperature exceeds about 90° F (32° C), evaporation saps heat away almost as fast as the Sun adds it. As a result, the Sun rarely warms the ocean past human body temperature. Even the most torrid tropical seas fall far short of the temperature you'd prefer in the shower. Truly hot water is rare in nature.

The most extreme exceptions are fissures deep in the ocean, where the planet's crust runs thin. Red-hot magma courses just beneath the ocean floor, superheating the ocean water in underground channels like the coils of an old radiator. Some of that scalding fluid, steeped with sulfur and metals, leaks through the ocean floor through small gaps in the crust. The water would boil if it weren't so deep and the pressure so high (see Chapter 4). It is in the hot, deep hydrothermal vents that ocean life has a chance to pit itself against the challenge of truly hot water.

But even in the comparatively balmy tropical oceans, or during warm days on cold shores, species can easily heat up beyond their tolerances. The oddest thing about warm ocean life is not just its heat-resistant adaptations, but also how many organisms live near the upper limits of their own physiologies. It is an odd principle of biology, but an Antarctic fish can die of heat stroke at about 43° F (6° C).[1] Corals live well at 81° F but suffer at 90° F (27° C and 32° C, respectively).[2] Extreme heat is almost entirely relative.

Alvin and the Pompeii worm

The three-person submersible *Alvin* silently descends through ever-darkening strata of water, on course for the absolute bottom: a hydrothermal vent 8,000 feet deep. When it arrives, powerful floodlights illuminate sulfurous chemicals pouring from holes in the sea bed. Thick black clouds belch out into the noxious water. Metal-rich deposits build up as the fluid rushes by, countless tiny flecks accreting over time into dark spires called black smokers spewing white-hot water out into the sea (see Chapter 4). It's the early 1980s, and black smokers are a recent discovery—made by past researchers piloting this very submersible.[3]

As the sub slows, its floodlights illuminate a lumpy and irregular 6-foot spire pouring black clouds like a Dr. Seuss smokestack. For a dozen feet in all directions, the floor practically explodes with life: tube worms, albino crabs, pallid shrimp, and other bottom dwellers—creatures already familiar to *Alvin's* crew. But as they pull in close, hovering in the still, benthic water, the scientists spy something novel on the smoker itself. Feathery red tufts adorn its lower reaches and when probed with a metal claw, they retract: they are in fact the fleshy, ornate heads of wriggling worms. Their rock burrows are piping-hot conduits for scalding water—but the worms are thriving. *Alvin* moves in for a closer look, and soon she's discovered another brand-new species. It's the Pompeii worm: the biggest heat-lover in the sea.[4]

Named for the submarine vessel that discovered them and an unfortunate Roman city suddenly buried in hot ash, *Alvinella pompejana* are found only at deep-sea hydrothermal vents. Delicate crimson plumes widen to fleshy gray bodies coated with hair-like bristles. The Pompeii worm's head sits awash in chilly 40° F (4° C) water, just a few inches outside the vent.[5] Its dense capillaries swell with dark red blood, rapidly exchanging oxygen with the water.[6] The tail, inches nearer the vent, may as well be on another planet. Subject to high and unpredictable temperatures, it's able to withstand water hotter than 120° F (50° C).[7]

Most animals do not live anywhere close to these temperatures. There is a thermal hot springs crustacean that lives about this hot, and some desert ants that live at 110° F (43° C).[8] But Pompeii worms hold the record for the hottest animal. They live so deep and in such unusual environments that scientists have only recently been able to secure a living specimen. Kept at 130° F (54° C), they die. So their hardiness, living on hydrothermal tubes where the interior parts of the worm tube reaches nearly 180° F (82° C), remains a mystery. Maybe Pompeii worms rapidly circulate fluids through their hot tails and cool tentacles, like a natural heat pump. Maybe the hot parts of the tail are adapted for high heat and the cold parts of the head are adapted for the cold. Maybe incubating the heads at 130° F is fatal, but the tails could take it. It would be an odd animal that had to build cells on one of its ends to withstand the hottest temperatures the ocean offers, and an inch away, had to build cells that functioned in the cold of the deep sea.

For this reason, molecular biologists took interest in *Alvinella*'s resilience from the moment the Pompeii worms were discovered. Their proteins and other cellular building bocks are some of the most heat resistant known in the animal world.[9] They could have any number of uses in both science and industry, so the race is on to gene-sequence both Pompeii worms and the similarly heat-resistant symbiotic bacteria that feed them.[10] Stout *Alvin*, still in service decades later, trucks down to the deep and gathers them into specialized high-pressure tanks in continued efforts to bring them to the lab.[11] Until we can regularly observe a living animal in a lab environment, most of the mysteries of the Pompeii worm from the ocean's burning pit will remain an enigma.

Deep see

The heat of the deep-sea vents, pumped out of Earth's bones, is a flickering flame in a vast night. The majority of the deep sea is dark and cold—ocean temperatures in the deep hover around 40° F (4° C).[12] The ocean-floor vents are tiny and rare, lined up along deep cracks in the ocean floor like rest stops on a dark desert highway, with many miles between them. Though they are extremely hot, vents don't cast heat far; even a few feet from a vigorous black smoker, the temperature drops from 650° F to 40° F (340° C to 4° C). Creep toward the smoker and get cooked; wander a few feet away, and the torpor of the dark and the cold settles in.

How does a vent animal tell where the black smoker is in a black ocean, where the only light might be brief bursts of bioluminescence? A person can feel a campfire's heat with eyes closed, and edge closer while keeping herself safe. But water absorbs heat, and heat receptors do not work well in water.[13] Enter the rift shrimp, a blind animal that sees hot water.

Rimicaris exoculata (literally, "rift-shrimp without eyes") is a cocktail shrimp–sized crustacean found exclusively near the smokers of hydrothermal vents. Between 2 and 3 inches long, covered in a transparent carapace, the rift shrimp spends its whole adult life at the edge of death, dancing across black smoker chimneys. Strong chitin toe-tips tear away gritty deposits from the smoker walls. The shrimp slurps down the sulfide-processing bacteria, and grinds the rest into a fine powder.[14] In addition, the shrimp have a whole community of bacteria living in their enlarged gills,[15] busily processing the hydrogen sulfide in hot seawater, like the symbionts in the hydrothermal tube worms (see Chapter 4). Either way, staying near the smoker is crucial, but the shrimp can't see it, lacking proper eyes.

Deep-sea biologist Cindy van Dover and her colleagues noticed two broad symmetrical patches of pigment on the back of these shrimp carrying heavy concentrations of rhodopsin: the same light-capturing pigment we have in our own eyes.[16] The patches have thin corneas and sensitive retinas, and optic nerves that connect directly with the back of the brain.[17] The patches are basically eyes—but not on the head. And the light that the shrimp sees with is not the light of the Sun—*Rimicaris* sees the eerie glow of the red-hot water.

When heated sufficiently, almost any substance radiates light in the low-frequency infrared spectrum. The exact frequency of light depends on the temperature—more heat, shorter wavelength and higher frequency. Our Sun emits yellow light because of its enormous temperature: 11,000° F (6,100° C). The heating elements of a red giant star, or a toaster, are cooler and emit more reddish light. The water of the deep-sea vents, at 650° F (340° C), emits a light at the very lowest end of red—just bright enough for the rift shrimp's eye patches to absorb.

Rimicaris has evolved the unique ability to actually perceive that light. The broad patches of rhodopsin on the shrimp's back are thought to increase the animal's ability to perceive dim light sources. There would be no room for such broad patches on the narrow eyestalks of most crustaceans. A reflective backing lies under the patches, acting like the reflective surfaces in a cat's eye to bounce what little light there is back up, should any fail to be absorbed the first time it passes through the rhodopsin patch.

The rhodopsin absorbs the light and conveys some sort of image directly to the brain. The glow is too feeble to form a true image, but rhodopsin can create a sense of proximity to heat. Although completely blind to light we'd call visible, his strange eyespots are perfectly built to perceive the only danger that really matters to him. Perceiving the vent's horizon by its own ember-glow light keeps him safe while mandibles skitter away at brittle sediment. Put a pillowcase over your head and walk into a lit room; you can perceive the light and shadows, but not see much else. Likewise, the rift shrimp perceive heat waves only vaguely.[18] They survive like the fictional blind samurai Zatoichi, tapping the ground with a walking stick while snatching arrows from the air.[19]

Corals, warmth, and death

Picture a secluded stretch of Samoan coast. Classic palm trees break up the sandy beach's blinding white, and behind them loom craggy volcanic hills. Warm, placid lagoon water laps at your ankles. A few hundred yards out in the water, tall rollers sputter their lives out in torrents of white foam. The submerged barriers sapping their strength can be seen from the beach: dark, amorphous forms. It's a coral reef, built over millennia into massive offshore walls. Coral polyps—tiny flower-like animals that clone themselves to form the living tissue of a coral—spread with an industrious passion unrivaled in the sea, every day secreting thin undercoats of enduring limestone.[20]

Over countless years, those microscopic films pile up into structures capable of feeding and sheltering thousands of other species. And even when the coral animals are dead and gone, the built-up limestone remains. The hard coral head cast up on a tropical beach is made of this limestone. Australia's Great Barrier Reef, one of the world's most spectacular natural wonders, is made of this limestone. Millions of years in the making, it is the only natural biological structure visible from space. The end result of industrious coral construction is the closest thing to a structured city you'll see beneath the waves.

Live reefs are magnificent to behold. Stunning blues and yellows and pinks and greens flash across the reef: fish, urchins, shrimp, and snails with a huge variety of body shapes and lifestyles. Worms huddle in their secret tubes, poking feathery heads out to troll the currents. Spires and walls of corals gradually rise from the ocean floor until the city hosts millions of creatures.[21] Even the white-sand beaches on tropical shores are mostly coral, ground to flecks over years by relentless waves and gnawing fish.

But for all their towering accomplishments, corals are disturbingly fragile. A temporary rise in temperature of only a few degrees can set off a major mortality event, wiping out whole swaths of polyps. Large cyclic heat waves in the Pacific, called El Niño weather, can devastate them. In 1998, this phenomenon killed up to 90% of the live polyps on some reefs.[22] In the past century, global warming from atmospheric carbon build-up has raised water temperature over a degree (Fahrenheit) in the tropics. That doesn't sound like much, but it spells trouble for such sensitive and foundational animals.

Corals feed themselves mostly by farming photosynthetic single-celled algae named *Symbiodinium*. Living inside the coral polyps' own body cells, they need plenty of sunlight and warm water to be productive. Thus, corals must live close to the surface in waters clear of cloudy sediment. These conditions are rarely found far from the equator, so that is where the overwhelming balance of corals live.[23] There is in fact a set distance from the equator—named to honor Charles Darwin's studies of coral reefs—beyond which corals can no longer build substantial reefs. The Darwin Point is located at about the latitude of Midway Atoll in the North Pacific, and just south of the Cook Islands in the South Pacific.[24]

Corals polyps can tolerate high temperatures, but their photosynthetic thralls are another story. Excess heat interferes with the algae's photosynthesis. The huge power of tropical sunlight, captured by the algae, is typically converted into high-energy electrons. At high temperatures, *Symbiodinium* leak these high-energy electrons like a pot boiling over. They bond into a nasty form of oxygen called *reactive oxygen species*. It's a toxin, poisoning the coral and forcing a response. The colony has only one option: expel the irritant.[25] The coral does this through the earliest evolutionary form of "cutting off your nose to spite your face"—by killing off its own cells. Sputtering algae are jettisoned wholesale into the sea. Most of the time the corals die and begin to crumble. This is called coral bleaching.

Predicting trouble for corals

In a profound discovery that is now used daily by reef health predictors, Paul Jokiel at the University of Hawaii's Institute of Marine Biology mapped the temperatures at which corals in different areas began to bleach. Along with his colleague Steve Coles, he noticed an oddly regular pattern: across the tropics, corals began to bleach at about 2–4° F hotter than the average annual peak.[26] Just a few weeks spent above that threshold triggered bleaching across hundreds of miles.

This data set was used by scientists from the National Oceanographic and Atmospheric Administration to devise a disarmingly simple metric: degree heating weeks. If temperatures persist 1° C (about 2° F) above bleaching temperature[27] for 1 week, the index goes up by 1.0. Two degrees (C) for a week adds 2.0, as does 1° C for 2 weeks. This simple index maps the danger to reefs from local warm spells, alerting locals and the world when a bleaching event is in progress. Nearly 90% of the reefs in Thailand suffered from bleaching in 2010, with up to 20% of total Thai corals left dead.[28] At this writing, there are not many patches of unusually warm water around the planet. The biggest one squats harmlessly north of Hawaii, just past the Darwin Point.[29]

The left coral branch from Ofu, Samoa, has bleached but the one on the right has not. They are the same species treated the same way, but the right coral was growing in a warmer Ofu pool and has higher heat tolerance. Photograph by Dan Griffin–GG Films.

Heat-resistant corals of Ofu

On the island of Ofu in American Samoa, the Sun rises at about 6 a.m. on the day before Christmas. It's a bright summer day in the southern hemisphere. Safe from the ocean swells behind a wall of coral, a lagoon sparkles in the sunrise like liquid cobalt. At dawn, the water temperature is already 84° F (29° C): a coral nirvana. By noon, fueled by the summer Sun and stilled by low tide, the lagoon reaches 95° F (35° C). The corals simmer for more than 3 hours at temperatures well past their normal tolerance. It's long past dusk when finally they cool below 90° F (32° C). Santa Claus is well on his way across the Pacific.

The corals of the Ofu lagoon should be dead from their daily bake, and yet they thrive. When their heat endurance was estimated in experimental hot-water tanks, Ofu corals were among the toughest ever tested.[30] New research shows that the hot-water pulses of the daily tides have sparked them to develop heat resistance. Twenty-four-hour exposure to 35° C heat would have killed them all long ago, but 3-hour stretches are bearable. In the classic film *The Princess Bride,* the hero has trained himself to resist poison in a similar way: gradual exposure, day by day, until once-lethal doses grow laughable.

Taking a page from human biomedical research, coral researchers measured how individual coral colonies use their genes during stress. Three days of heating activates a battery of 250 different stress genes in the typical coral. In the Ofu lagoon, the corals keep about 60 of these "heat genes" operating at high capacity all the time. Some of these corals seem to be born with these guardian genes turned on, but others only turn them on when moved by scientists to the reef's hottest region. Some never activate the crucial genes; these colonies simply die. The cumulative result is a small band of survivors thriving in a small backreef lagoon a quarter mile across, growing in the intense sun and heat. Though they are the Pacfic's toughest known corals, human interference still pressures them. Overfishing leads to choking algae growth, a landfill leaks heavy metals into the lagoon, and some on the island look to improve commerce by extending their tiny airstrip directly out over the reef, right where these corals live.

Luckily, these reefs are partly protected and deeply appreciated by the local villages, which are caught between their conservationist impulses and the realities of economic development. We still have time to figure out the survival secrets of the mighty coral polyp, to help the villages of Ofu protect their reefs, and to see whether their coral's survival skills can be duplicated.

The red hot sea

Nestled near the Great Rift Valley of Africa, where some of the planet's tectonic plates come together, lies an unlikely place full of corals. The Red Sea is deep, plunging down into a 7,000-foot trough, but the shallow edges of this desert ocean are ringed with spectacular reef formations.[31] The water is extremely warm. Fueled by the murderous Middle Eastern Sun and the surrounding desert, the average summer water temperature is 86–88° F (30–31° C), high enough to bleach most Pacific corals.[32]

From the air, the Red Sea coast transitions from white sandy beaches, to sapphire blue water, and finally to a convoluted brown crust like a million muddy cul-de-sacs. Red Sea corals form "fringing reefs," growing out in broad sheets toward the steep shelf of the coastline.[33] Charles Darwin himself took an interest in the Red Sea's corals, describing their distinctive fringe formations. But some coral structures made no sense to him: they stood far apart from the land, seeming to describe the curves of coastlines that didn't exist. Others were isolated columns of ancient limestone, thrust up from the ocean floor in irregular spires teeming with colorful fish. Darwin was at a loss to explain them:

in the Red Sea, and within some parts of the East Indian Archipelago (if the imperfect charts of the latter can be trusted), there are many scattered reefs, of small size, represented in the chart by mere dots, which rise out of deep water: these cannot be arranged under either of the three classes: in the Red Sea, however, some of these little reefs, from their position, seem once to have formed parts of a continuous barrier.[34]

As it turns out, these formations were lingering evidence of a turbulent and unusual history. The Great Rift Valley is in a state of constant geologic change as its three plates struggle to pull away from one another.[35] The competing forces make the Valley a giant seismic anomaly: a never-ending war between three drifting continents that raises and lowers the level of the Red Sea. At this moment, the level of the Red Sea is at a historical high: about 150 feet higher than the half-million-year average. The spires of coral rock and offshore reefs that Darwin puzzled over are flooded now but were in a normal position along the coast during those older epochs.

The hot, salty Red Sea roller coaster might seem an inhospitable place for glacially slow-building coral reefs. Yet marine species thrive there, feeding off opportunistic corals and evolving rapidly into new species. Of all the reef fish in the Red Sea, 10% are found there and nowhere else.[36] Many changes are purely cosmetic. The crown butterflyfish, *Chaetodon paucifasciatus,* is a stunning fish with black chevrons running lengthwise down its sides and a black, yellow, and white bar running through the eye. A broad crimson patch is splashed on the abdomen, mirrored by a similar colored band on the tail.[37] This species is found only in the Red Sea, but its nearest relative (according to DNA evidence) is an almost identical species in Madagascar.[38] Its patches are duller, demoted from scarlet to mustard yellow.

The local corals are similar to those found elsewhere but have evolved some unique abilities. Red Sea reefs, living close to their bleaching temperatures, contain heat-resistant symbiont algae. They're a type of *Symbiodinium,* the common coral symbiont, yet they're able to keep their hosts healthy even in much warmer water than usual. How this happens is not completely known. But there's little doubt that the Red Sea *Symbiodinium* gives these corals an edge during the sweltering summer heat. And across the Red Sea, the hotter the water, the more common the hot symbionts become.[39]

Red Sea corals seldom bleach in the normal range of summer temperatures.[40] But like so many of their cousins, they have very precise thresholds for survival. Heat-resistant algae have their limits. Anne Cohen and her colleagues at Woods Hole Oceanographic Institute measured the past growth rates of corals in a CT scanner: a version of the medical equipment physicians use to measure human bone growth.

Coral biologists can measure growth rings in coral heads the way foresters read tree rings. Cohen and her colleagues found that in the 1940s (when there were two consecutive hot 90° F [32°C] summers), corals grew at just a fraction of their expected rates.[41] Recent temperatures in the Red Sea are ticking up; its water warms with the rest of the globe. As Cohen's research suggested, coral growth is already trending down. Red Sea corals may be the world's toughest, but they're strained to their limits.

Vaquita

The shoreline in north Baja is a bleached mosaic of old boats: simple skiffs with outboard motors, green and blue and pink that used to be red, all crammed with fishing gear and white gill nets like billowing clouds. The water is shallow, minestrone-warm, and bears the chromatic sheen of leaked boat fuel. Fifty yards out, the harbor is only knee-deep. At its edge, where the water turns from sandy blonde to topaz, sea turtles pop up their heads for bubbling breaths. Tall formations of white stone loom like ivory castles, guarded over by rowdy seabirds. This is the northern tip of the Gulf of California: the world's warmest open waters.[42] Their soupy productivity sustains masses of shrimp and fish, jumbo squid, summering whales, and some of the ocean's most diverse microorganisms. They also harbor a unique species of mammal on the very edge of extinction: vaquita, "little cow," the pygmy desert porpoise.

The vaquita, *Phocoena sinus,* is the smallest and most endangered cetacean on the planet, a dolphin-like creature just 5 feet in length.[43] Rubbery skin covers the little cow in a solid gray raincoat, but dark markings ring her eyes like a bandit's and lend her lips the appearance of a coy smile. Vaquitas inhabit a tiny triangle of shallow water, 40 miles on each side and wedged into the Sea of Cortez's northernmost tip. They live nowhere else in the world, in climes warmer than those inhabited by any other cetacean.[44]

Living in the northern Gulf of California is like being chained in a hot tub: pleasant at first, a trial over any duration. Mammals can tolerate extreme heat for short spans, but no marine mammal encounters the kind of heat the vaquita endures all its life.[45] Water temperature spike high in the summer, as

Vaquita porpoise statue in San Felipe, Baja California. Photograph by Cheryl Butner.

hot as anything in the Red Sea, and a warm-blooded mammal struggles to shed its own body heat. Small terrestrial mammals have a high surface-area-to-body-mass ratio and rarely overheat. Large mammals have contrived tactics like sweating, panting, or ear flapping to cool off. But vaquitas are caught in a bind: large enough to struggle with heat but trapped in water, where they can neither sweat nor pant.

To compensate, the little cows have two adaptations. First, they've shed their blubber. Vaquitas don't look skinny compared to typical porpoises, but they carry far less body fat. Dropping insulation keeps heat from building up.[46] Second, their fins are much larger than a typical porpoise's. From dorsal to pectoral fin and back to the tail flukes, these oversized appendages act as radiators.

"Little cows" have tough lives, but the heat doesn't kill them: no deaths by heat stroke or summer fever have been recorded. They seem to have adapted to this environment and specialized to it - but their triangle of hot water is so small that they are one of the rarest mammals on Earth. Their first recorded species census (they were only recognized as a species by Ken Norris and William MacFarland in 1958) posited less than 600 individuals.[47] They are only one of six "true" porpoise species in the world, and their nearest relatives are the South American coastal Burmeister's porpoise and the spectacled porpoise of the Antarctic.[48] Both these species live in much colder waters. Neither lives within 1,000 miles of the vaquita. How and when did the pygmy porpoise reach the Gulf of California and evolve into a hot-water specialist?

DNA evidence suggests that vaquitas entered the Gulf of California 2–3 million years ago: a turbulent time amidst the Pleistocene Ice Ages, when sea surface temperatures varied widely. With massive glaciers locking up much of the planet's water, and the cold of the poles reaching much farther toward the equator, the typical girdle of warm water in the tropics narrowed. Perhaps a roving pod of porpoises crossed the equator at that time, seeking greener pastures like the first ancient caribou to cross Alaska's Bering Strait. Once in the Gulf of California, they may have been trapped by some accident of geography or climate. However it happened, the vaquitas' survival proves it made the needed evolutionary changes. Yet those very adaptations make it difficult for them to live anywhere else.

Close to boiling

Pompeii worms live through scalding black-smoker belches of 150° F (66° C), but a typical Antarctic fish dies of heat stroke at 43° F (6° C), what we'd call ice water.[49] How can species have such different sensitivities to heat? Digging down to the level of genomes, there are lots of tricks evolution can play to adjust physiology to higher heat. For instance, the amino acids in proteins can be re-engineered to retain their shapes even at high temperature. Work on the Pompeii worm shows that wholesale evolution can take place across the genome to craft cells that function in hot-vent habitats.[50] Species that control their body temperatures, such as mammals and birds, often evolve changes in size or metabolic rate. And there are many ways for an individual to adjust its internal physiology to accommodate temporary heat stress.

In the face of these mechanisms, it is perhaps a surprise that so many species are so sensitive to small increases in water temperature. Biologist Jonathan Stillman made a careful study of the heat tolerance in coastal crabs, from the hot shores of the Gulf of California to the chilly fog of Monterey.[51]

Depending on where a species lived, he observed heat-induced heart attacks at just a few degrees above their environments' typical peak. Tropical crabs, the ones on the hottest shores, had less leeway than their temperate cousins. Even though they had more tolerance of average heat levels, their heat ceiling was just a few degrees above normal. A few unusually hot days could kill them.

Stillman's discovery changes how we have to think about future warming, because it means that species that are adapted to high temperatures are *not* more likely to be safe when future oceans heat up. Our normal thinking would intuitively suggest, "They are already good at tolerating heat, so they should be able to survive some extra warming." But when heat-resistant species are already living at the upper edge of their tolerances, they are no more able than other species to survive one extra degree.

CHAPTER 9 THE COLDEST

Cold water is often very productive,
but ice is a danger inside and out.

Unicorns of the sea

Picture a fourteenth-century medieval castle in the grip of a sweltering summer heat wave. The thick stone walls of the keep are cool, but the air is humid, and the walls run with dew. In the audience chamber, bright rays of sunlight pour through thick windows set high in the wall. A retinue of guests present themselves to the regent, offering cool exotic spirits as a gift for his hospitality. The prince accepts and calls for a table to be set. Stewards lay out goblets for the visitors, but a smaller vessel is set before the throne: narrow and fluted, bone-white and carved with equine forms. The foreigners narrow their eyes; some conceit to set the prince apart from his guests? His drink poured, he takes the cup and rises. They toast and sip and sit together. *If it doesn't offend, my Lord, what is that cup?* The prince smiles in response, holding it up to the light.

"This is carved," he intones with a pause for effect, "from the horn of a unicorn, hunted by Norsemen in the distant north. You know its properties?" They do, but he continues: "Among others, the immediate and absolute negation of any poison. I'm a trusting man," he looks at the visitors around him, "but these are dangerous times we live in."

These details are fiction, but the circumstances aren't. Centuries ago, a highly lucrative market for unicorn horns existed across Europe. Traveling merchants brought them from the north: fine spars of braided ivory several feet in length. Legend held these horns were powerful antidotes to both poison and witchcraft; rulers across the continent paid more than the artifacts' weight in gold.[1] Queen Elizabeth I paid the price of a castle for a single jewel-encrusted horn.[2]

It should go without saying that unicorns have never existed. But those horns came from *somewhere*, and they really were harvested by Norsemen. The creatures they hunted weren't found on land, but rather in the freezing black waters of the Arctic Sea. They hunted the small cetaceans known as narwhals.

The name derives from the Old Norse *nar,* meaning "corpse."[3] The Norse were the first Europeans to sight these toothed whales, and their mottled gray bodies hanging near the surface may have looked like dead men floating in the water. Narwhals, *Monodon monoceros,* congregate in pods of several dozen, churning up the surface between rafts of pack ice, feeding on fish and squid. The biggest populations are in the Eastern Canadian Arctic and Greenland, and narwhals feature prominently in the lore of northern Canada's Inuit natives: the first humans to encounter and hunt them.[4]

"Unicorn" horns became staples of Medieval European collectors, earning pride of place in cabinets of curiosities.[5] That "horn" is really a tusk: a supremely elongated incisor protruding from the upper left side of a male's jaw. Females don't usually have tusks, though rare exceptions exist. Even rarer are the double-tusked males, with mighty spars erupting from both sides of their overcrowded mouths. The average tusk is about 8 feet long, attached to a 15-foot male.[6] This is ungainly even in the sea. On land, a unicorn so proportioned would simply fall on its majestic nose.

The regents of Europe prized the tusk for its aesthetic splendor, and narwhals seem to feel the same. It doesn't aid survival; tuskless females lead perfectly healthy lives and, for the most part, behave similarly to males. Rather, the tusk is a secondary sex characteristic, deployed in the service of mating. Males use their tusks to compete for females and can sometimes be seen locking horns. They clash with the awkwardness of dueling marionettes— imagine fighting another person while you both hold broom handles in your mouths. This behavior is more akin to posturing than violence; an arm-wrestling match, not a brawl. It seems to be the tusk's only practical function. Though males have been observed rooting in the muddy bottom for food,[7] a primary, practical purpose for the tusk has yet to be determined.[8]

Narwhals are a migratory species in some of the world's coldest seas. During the Arctic Ocean summer, they can be found cavorting in great shoals of stone-gray flesh. Most polar species eat heavily during summer, when the cold, nutrient-rich seas and almost constant sunlight create a biological boomtown—socking away calories for the winter. Narwhals do the opposite: diet studies in summer repeatedly turn up animals with empty stomachs.[9] Feeding happens during the winter, when narwhals retreat south to the open

The "Unicorn sword" in the Imperial Treasury (Schatzkammer), Vienna. From Pluskowski, A. 2004. "Narwhals or unicorns? Exotic animals as material culture in Medieval Europe." *European Journal of Archaeology* 7:291–313. © Kunsthistorisches Museum Wien.

ocean. Even during the year's coldest months, they remain stubbornly north of the Arctic circle. This southern migration frees them from the worst seasonal pack ice, which always threatens to wall them up in coastal inlets.[10] Once in the wide deep sea, they dive a half-mile deep to gorge on a bottom-dwelling buffet of cod, halibut, and shrimp. A wintering narwhal accrues massive deep-water mileage, spending hours each day on repeated deep dives, inhaling seafood to make up the calories.[11]

Narwhals have evolved—biology and behavior—to icy surroundings, but they started out in warmer water. The fossil record shows that their ancestors swam in temperate waters some 3 million years ago.[12] Modern narwhals and their beluga cousins are confined to the Arctic, their climate adaptations an eerily mirrored image of those seen in the Gulf of California's vaquita porpoises (see Chapter 8). Immersed in bathwater warmth, vaquitas are small, with outsized pectoral and dorsal and fins to radiate away excess body heat. Narwhals take the opposite approach, weighing in heavy for porpoises, sporting small pectorals and totally lacking dorsal fins. This helps conserves heat and may help the whales negotiate tiny openings in the Arctic pack ice. Even in summer, unpredictable ice floes might crowd the surface and make it hard to breathe.

Both narwhals and their beluga cousins carry advanced echolocation organs, from soft "melons" at the front of their skulls to fat-filled cavities in their lower jaws. These enhanced organs focus outgoing sound pulses and perceive incoming reflections. Perhaps the narwhal's oversized equipment finds food in the dark deep oceans, or perhaps it maps the local pack ice in fine detail. Accurate echolocation could tell whales which paths dead-end and which might lead to a precious gasp of air.[13] A typical whale operates in an open-water environment, where obstructions are rare. Arctic whales demand much more from their sonars.

Twinned reasons sea otters survive the cold

When discussing the ocean's coldest residents, the word "cute" rarely comes up. Polar bears might be terrifying, beluga whales merry, haddock hardy like grizzled sea captains; but extreme animals are seldom cute. Enter the sea otter, *Enhydra lutris*—cute enough to inspire Internet memes and stare preciously from countless wildlife calendars. Nearly every American schoolchild can picture it: lustrous coat, dark eyes gleaming with a bright inquisitive intelligence, floating placidly on its back while eating off its belly. Conservationists and marketing executives alike bend their knees to the mighty sea otter.

The first people to profit from the otter were after its pelt. Sea otters' amazingly warm, fine coats are their most valuable adaptation, enabling survival in truly frigid water. Living exclusively in the cold North Pacific, three subspecies of sea otter spread over millennia from the northern shores of Russia across the Bering Strait and down North America's west coast.[14] The California Current drifts eternally south, bringing nutrient-rich water with it and powering massive productivity. Giant kelp—the world's largest algae—can grow a foot a day off this frigid stew.[15] Sea otters live among the kelp forests in water ranging from 45° to 60° F (7° to 15° C). There are dozens of marine mammal species living nearer the poles than this, but otters are unique: having evolved from river otters, they're small and lack blubber.

Large size and thick insulating body fat (blubber) are the gold standards of cold-water survival. Whales and seals would freeze to death without these adaptations. Otters are the smallest marine mammal—about 60 pounds and the size of a dog—yet do entirely without blubber, relying solely on their magnificent pelts for insulation. A single square inch of otter fur sprouts up to a million filaments of hair, distributed into two layers.[16] The fur is downy to the touch, smooth as silk and practically radiating warmth. A short oily undercoat is protected by longer, fibrous guard hairs, which trap a paper-thin membrane of air in the fur as the animal dives. Air is an excellent insulator, and the oil of the undercoat ensures water is kept away from the otter's skin. The air bubble is actually visible as an otter swims: a shimmering silver cloak that warps and twists with the otter's lithe movements.[17] Countless mammals rely on heavy coats for warmth, but otter fur is in a class of its own.

Ironically, that wonderful fur nearly drove them to extinction. When European explorers discovered them the middle of the eighteenth century, a worldwide market swiftly erupted. Wealthy people from China to Europe dressed themselves richly and warmly at otters' expense. North American sea otters were hunted to the brink of extinction in just 50 years: an estimated 100,000 otters were killed from Northern California to Baja.[18] Still more were taken by Russians along the Alaskan coast. By 1844, otter numbers were so depleted that the governor of California (then a part of Mexico) issued North America's first fishery-conservation decree: no juvenile otters were to be hunted.[19]

Sea otters boast a second adaptation to answer the ocean's seeping cold, one that combines with the first to keep them alive in the North Pacific. Their bodies are raging metabolic furnaces. Constant immersion in cold water is a massive energy tax, and otters have evolved an industrious metabolism to pay it. Of course the furnaces need fuel, so every day a 60-pound otter eats a quarter of its body weight in fresh seafood.[20] If it doesn't eat for a day, it loses weight at a rapid clip. Just 3 days of starvation can kill an otter, freezing it to death. A healthy otter population consumes a staggering amount of ocean life, from sea urchins and abalone to crabs, snails, and clams. Virtually anything an otter can bite into or smash on its belly with a rock will end up in its stomach. That appetite and flexible diet has made the otter a linchpin in the ecology of the kelp forest.

Urchins and abalone are otters' favorite meals, subjected to such voracious predation that they rarely become too abundant. But once the sea otters were wiped out (or near to it), these ocean herbivores ran amok like locust swarms. The West Coast's great kelp forests were mowed down by an army of echinoderms leaving nothing in the forests' wake but fields of barren rocks and basketball-sized sea urchins. Aerial photographs of the Monterey coast in 1906—perhaps 60 years after anyone saw an otter there—show none of the thick kelp forests that line the coast today. Divers in the early twentieth century reported acres of sea bottom literally carpeted with the urchins, purple and red. Only a few tenacious strands of brown bull kelp existed: a smaller, less productive species.

Despite the extensive fur trade, sea otters managed to survive in small bands. Alaska's remote Aleutian Islands harbored a handful.[21] Otters had also survived in pockets along California's rugged and undeveloped Big Sur coast, protected by local groups such as the Friends of the Sea Otter.[22] With years of careful shelter, otter populations made a slow comeback. Being cute certainly didn't hurt, and their "rediscovery" in California was a public sensation.[23] In April 1963, sea otters returned to Monterey Bay for the first time in perhaps a century. Tourists and locals lined the rocky shores at Lover's Point, eager to see these enigmatic creatures. Today they remain one of the Monterey Bay Aquarium's most popular attractions.

The kelp forest expanded in the wake of the otter population and the decimation of the urchins and abalone. Graduate students at the Hopkins Marine Station followed their progress up the coast from Big Sur to Monterey, noting an ocean floor littered with broken sea urchin skeletons like some bizarre alien boneyard.[24] A similar chain of events happened in Alaska: where the otters went, the urchins disappeared. With the urchins gone, kelp bloomed in their wake.[25] Today the kelp forest supports vast populations of fish, seabirds, and seals. Its productivity sustains the livelihoods of countless people along the Pacific coast. The peculiar cold-water adaptations of sea otters make them crucial to the kelp forests' health. Their revival is the best ecological news in a century for the North American West Coast.

Antarctic antifreeze

In the winter, Antarctic Ocean temperatures may drop to 28° F (−2° C). That's well below the freezing temperature of fresh water, and even in salt water ice crystals start appearing.[26] Blood and bodily fluids are less salty than seawater,

Blackfin icefish, *Chaenocephalus aceratus*. Photograph by William Detrich, U.S. Antarctic Program.

so ice appears more readily inside a fish than outside it. For living cells, ice is death: a sharp crystal scimitar slicing through membranes. At a larger scale, ice might lodge in small capillaries and trigger a stroke in the brain. Polar fish must work constantly against ice formation inside their bodies.

When the Antarctic Ocean's seasonal freezes first began some 10–14 million years ago, its residents were forced to adapt.[27] Among the most successful were the Antarctic fish classified as notothenioids. Collectively known as icefish, these thin, large-eyed little creatures have prominent lips, making them look like mustachioed RAF pilots. They were the first to evolve a remarkable set of genes used to build an amazing natural antifreeze protein (AFP).[28] AFPs have two functions. First, they alter their internal environments, decreasing the freezing point of blood so that ice crystals cannot form.[29] Second, if ice crystals should appear, they're swiftly gummed up by AFPs binding directly to their open faces. This is a direct intervention, physically interfering with change in the ice crystals. They cannot easily expand, melt, or refreeze.[30] The end result is a safe internal environment, where ice crystals are both rare and stable.

University of Auckland biologist Clive Evans and colleagues looked at this, asking the obvious question. Since antifreeze proteins—especially the sugar-coated variants called antifreeze glycoproteins—can't totally eradicate ice crystals, what happens to the particles that remain?[31] Evans coated tiny particles with antifreeze glycoproteins, injected them into polar fish, and watched them accumulate in a particular organ. Icefish spleens have a unique function: identifying ice crystals wrapped in glycoproteins and storing them.[32] The proteins patrol the fish's bloodstream, moving from the digestive tract to the blood supply. They snap up ice crystals, swaddle them in protective chemicals, and deposit them safely in the spleen.

Antifreeze is obviously a major evolutionary advantage: in the Antarctic Ocean, icefish and their relatives employing these proteins account for 95% of the region's fish biomass.[33] At the planet's opposite pole, eelpout and Arctic cod have independently adapted nearly identical proteins to survive farther north (beyond 84°N—the North Pole is at 90°N) than any other fish.[34] Antifreeze proteins also appear on land, mostly in cold-adapted insects and plants. Though similar compounds have evolved many times in many places, they operate in the same way: threonine amino acids binding to the flat surfaces of ice crystals like wet tongues to cube cubes. The crystals don't melt, but they're manageable so long as they can't grow. Antifreeze proteins are so useful that the Protein Data Bank once awarded them "Molecule of the Month" status.[35] The little atom clusters were reportedly embarrassed by the honor.

Human chemists have designed a whole constellation of products based on fish proteins. AFPs are highly effective and nontoxic, whereas conventional chemicals like ethylene glycol (vehicular antifreeze) are lethal poisons. You can buy heat-resistant ice cream at many grocers, laced with AFPs to keep it from too easily melting or forming overlarge ice crystals.[36]

The AFPs in commercial foods got there the hard way: icefish genes were inserted into yeast cells, and the yeast produced ice-suppressing proteins using their new genes. The resulting proteins bind to ice crystals in ice cream and coat them. The end result tastes creamier and prevents large crystals from growing in a frost-free freezer. It's lower in fat, cholesterol, and calories while still retaining the taste of a less-healthy treat.[37] The technology to enhance our ice cream evolved in pioneer fish, inhabiting the ocean's coldest climes.

Antarctic krill

The Southern Ocean flows around and around Antarctica in a constant clockwise current, blown by the west wind.[38] It also flows slightly north, a spiral careening outward, taking the sea's surface layers with it. Those layers are replaced by nutrient-rich water rising from the ocean's bottom—bitter cold but still warmer than the icy surface. This upwelling constantly transfers nutrients from the bottom of the sea (where they naturally accumulate) to the top.

When the pack ice retreats in the spring, sunlight returns to the polar sea. Its warmth and energy combine with the nutrients for a bonanza of plank-

The krill *Euphausia superba* is the base of many Antarctic food chains. Photograph by Uwe Kils.

tonic growth.[39] The result is magnificent: a huge explosion of tiny emerald cells drinking in the Sun's bounty. By January (summer in the Southern Hemisphere), the plankton bloom attracts countless grazers. The most important are the thumb-sized crustaceans called Antarctic krill, *Euphausia superba:* one of the world's most successful species.

It depends, of course, on how one defines success. *Homo sapiens* have accomplished a great deal in a quarter-million years. But in terms of gross mass, krill take the heavyweight belt—the combined biomass of the world's Antarctic krill hovers around 350 million tons.[40] There are an estimated 800 trillion of them—outnumbering human beings 100,000 times over. These busy little creatures crowd to the surface in swarms so dense they beggar belief, clawing their way up to snatch specks of floating algae. Their translucent shells are tinted red and studded with green bioluminescent lights like tiny LEDs. The thickest swarms contain 30,000 individuals per cubic yard, roiling in giant pink clouds so dense they look to be solid objects. During the summer boom, a single cloud may stretch for miles and comprise billions of individuals.[41]

These animals are hardy, tenacious survivors. They endure the winter and spring freezes when pack ice plates the ocean over, and have the ability—

unusual among shelled organisms—to grow smaller with each molt rather than larger. During the leanest months, they metabolize their own tissues and wither away beneath the ice. In lab settings, Antarctic krill have survived more than 200 days without food.[42] But it wasn't until 1986 that scientists discovered where this hugely abundant species spent its winters. They persisted beneath the sea ice, clinging upside-down to frozen ceilings and enjoying a food supply nobody imagined: they fed on green algae growing on under surfaces of the ice.[43] Adults live 2–3 three years, feeding on their ice farms each winter, waiting for the Antarctic night to give rise to the unending Antarctic day.

Krill are eaten by nearly everything larger than themselves, from fish to penguins to the great whales. They're the easiest and most plentiful meal in the Southern Ocean, so abundant that many species specialize on them. The misnamed crabeater seals pursue krill rather than crabs, having adapted their teeth into bizarre clover-like cusps that act as sieves for the pink shrimplike morsels.[44] Whale baleen is a similarly peculiar adaptation, designed to filter out krill from a large mouthful of water. Where krill are concerned, eating a lot of them efficiently is the battle plan for big marine mammals.

Krill are a shortcut in the food chain. In a typical ecosystem, the tiny algae called phytoplankton and high-order predators are distantly linked: sunlight powers phytoplankton, which are eaten by grazers, which are eaten by small predators, and so on until we reach apex predators like sharks. Enormous quantities of energy are lost at each level. As a general rule, organisms pass on only 10% of their bodies' total food energy when consumed—10 pounds of phytoplankton produces 1 pound of grazer.[45] Krill cut the food chain short, transferring energy from the limitless Sun to great whales (and everything else) in just two steps. Because of this, the Antarctic ecosystem is outrageously efficient. But even something as vast as the krill population is not too big for humans to affect. We struck the first blow inadvertently, by hunting krill predators.

Thar she goes

Whaling has been a human occupation for thousands of years, yet it was only in the twentieth century that the Antarctic's fertile whaling grounds were finally tapped.[46] Despite their late start, hunters fell on the whales of the Antarctic like ravenous hyenas. From 1907 to 1985, humans killed more than a million blue, fin, humpback, and sei whales in the Southern Ocean.[47] These whales ate countless truckloads of krill every year, and so many were taken that a krill surplus may well have emerged in their absence.[48] In a self-serving extension of this hypothesis, the Japanese Foreign Ministry has contended that the killing of great whales led to an overabundance of small baleen whales like the minke (spared from major hunting until the 1980s). Given this assumption, they assert that "culling of minke whales may greatly help the recovery of the larger baleen whales."[49] If true—if great whale populations really are constrained by krill stocks—this would provide a scientific justification for the continued Japanese hunting of Antarctic minkes.

Until recently, the claim that minke whales (genus *Balaenoptera*) were overpopulated weeds (or "cockroaches of the sea" in the words of one Japan's former Whaling Commissioners) could not be directly tested.[50] No hard population numbers exist before the twentieth century. In 2010, a new kind of DNA analysis refuted the "cockroach" claim, providing a fresh glimpse into the Southern Ocean's ecological history. The technique is based on the tendency of bigger populations to contain higher levels of genetic variation than smaller ones do: by measuring present variation, we can estimate past population sizes.

When this approach was applied to Antarctic minke meat—ironically, purchased from the Japanese whaling industry—the results put the lie to the Japanese whalers' claims. There were no more minke whales in 2010 than before whaling's Golden Age: about 700,000 animals.[51] Without evidence of recent overpopulation, there's no need to kill minkes.[52] The Krill Surplus Hypothesis remains reasonable, but it does not explain the abundance of minke whales or evidence for a recent decline over the past few decades. It's possible minke numbers are determined by winter mortality or the loss of sea ice over the past 50 years.[53] Either way, there is no scientific basis for culling minke whales.

Energy from the cold

Overexploitation of the sea is an ongoing crisis. Between fishing and pollution, there can be no doubt *Homo sapiens* have seriously compromised the ocean's capacity. But what if the ocean's most powerful engine could be harnessed for our own uses? The incredible productivity of the Antarctic environment—from phytoplankton to krill and on down the line—is the result of cold nutrient-rich water rising from the bottom to the surface. If we imagine this cold water as a resource in its own right, a light bulb suddenly winks on.

There are only so many fish in the world, all reliant on delicately balanced eco-systems and susceptible to countless influences. In contrast, the planet's sup-ply of cold seawater is essentially infinite and incorruptible. Pumping huge volumes of heavy water from great depths to the surface is a simple engineer-ing problem, but the energy costs are high. How could anyone turn a profit doing this?

As far back as the late 1800s, physicists sought to tap this energy source. Using the difference in temperatures between warm surface water and cold deep water to propel a motor via crude thermodynamics, they could gener-ate heat and steam and, ultimately, electricity. Georges Claude built the first of these plants in Cuba in 1930, powering a turbine that lit forty 500-watt bulbs.[54] The system was small and highly inefficient—barely producing more energy than it consumed—but the principle was sound. They called it ocean thermal energy conversion, or OTEC. Work continued apace, and by the late twentieth century, the modern technology existed to escalate this old idea.

In 1974, the U.S. government established the Natural Energy Laboratory of Hawaii Authority (NELHA). Sited at Keahole Point, Hawaii, on a spit of barren black lava stretching out from the dry Kona coast, the facility resem-bles a mad scientist's compound. Satellite dishes grasp at the sky; mysterious concrete domes sit beside laboratory buildings of gleaming aluminum and blue glass. OTEC is feasible only in the tropics, where bathwater-warm sur-face temperatures quickly give way to deep cold currents. The greater the dis-parity in temperature, the more electricity can be wrung from the gradient. Keahole Point is special: formed by a relatively recent lava flow, it juts out like a long pier over a sharp continental shelf. Jagged basalt ends abruptly at the water's edge, where ocean swells slosh awkwardly without a shallow bottom to trip them up. Just a few yards offshore, the ocean is already more than 300 feet deep.[55] This fortuitous undersea cliff was the perfect proving ground for experiments in renewable energy.[56]

Like most things in life, the results were neither so good as hoped nor as bad as feared. Pumping so much water takes a lot of energy, so nearly all the power OTEC produces must be cycled back into running the plant. Once operational, the Keahole plant produced more electricity than it con-sumed but not enough to be economical, even with the inflated energy prices endemic to tropical islands. NELHA was in danger of failing. The solution ulti-mately came from the water itself; the plant used its temperature gradient but neglected the nutrients stored within. Those nutrients could be used for cultivation!

The term "artificial upwelling" was coined by Oswald Roels in the 1970s,[57] as he grew algae, shrimp, and shellfish in the Virgin Islands using deep water pumped to the surface.[58] Cold, rich seawater from offshore sites yielded high growth potential, and aquaculture might defray the high pumping costs. The same approach was quickly adopted in Hawaii in 1983 once it became apparent that NELHA couldn't survive solely as an electrical supplier. High-value marine species were tested and grown: the ubiquitous Japanese seaweed *nori* grew 30% each day, and California kelp grew quickly enough to feed cold-water abalone thousands of miles from their natural habitat. Even the last depleted dregs of used seawater could be used to grow high-nitrogen *Spirulina* algae—an important component in both animal feeds and human dietary supplements.[59]

No single project at Keahole Point has a large output, and none could stand on its own economically. But the net result—electricity, agriculture, and aquaculture supplied by an infinitely renewable resource—has a much better chance.[60] In a world with too much atmospheric carbon and too few fossil fuels, OTEC production might be a key to stabilizing tropical island economies under siege from rising seas and high prices.

Glass sponges

Most corals can't survive the polar seas. The ones that do are slow-growing members of the black coral family, living fragile and solitary lives anchored to rock promontories (see Chapter 6); or the solitary corals, a grab bag of different corals that have lost their colonies and live as single polyps; or anemones that have invented skeletons.[61] Cold-water reefs do exist in very deep places, but the organisms aren't corals. Rather, they're hexactinellids: bizarre creatures closely related to sponges, made of living glass.

Colloquially they're known as glass sponges, sharing ancient origins with the conventional sponges that fill every ocean on the planet and titter obnoxiously on children's television.[62] Unlike those poorly organized smears of tissue, glass sponges are enterprising builders. Collecting silicon dioxide from the surrounding water, they assemble it into six-rayed crystalline forms like toy jacks. These tiny crystals, called spicules, become the building blocks for some truly spectacular construction. The sponge stacks the crystals and grows around them like ivy on a trestle, armoring up its own soft tissue with a brittle exoskeleton. Tube-like spires of white silica emerge from the ocean bottom, growing like living icicles.[63]

Glass sponges are crucial anchors of cold-water communities, taking their social responsibilities seriously in a sea too cold for corals. Off the northwest coast of North America lie deep, dark canyons. Sheared away by glaciers like the mighty cliffs of Yosemite, the last Ice Age sunk them into icy tombs. At the bottom sit the some of the largest structures ever built by living things.[64] When glass sponges die, their flesh decays, but their skeletons remain. Generation after generation of sponges slaved away in the cold off British Columbia, piling one skeleton atop the last. After thousands of years, the results are breathtaking: vast reefs, hundreds of yards wide and dozens thick, stretching for 25 miles down silent corridors along the coast.[65] They teem with life; fish and tiny shrimp flit between the sponges' sunflower-yellow fingers, eating and thriving despite the freezing water. Biologists speculate that in ages past, the whole North Pacific was striped with these reefs. The titanic structures thrived before tectonic action and warming seas laid them to waste: cathedrals of spun glass, shattered and sprawled like the remains of fallen gods.

A few hexactinellid species mimic true corals by farming their own vegetable gardens, in the form of green photosynthesizing algae. The algae live clustered around the glass skeleton, transferring their energy surpluses to the host sponge. When researchers discovered this phenomenon, they were confused. How could algae possibly obtain light so deep under water and so far inside the sponges' tissue? As it turns out, the glass skeleton transmits light. Just as fiber-optic cables use tiny glass filaments to transmit information, glass sponge spicules gather the feeble sunlight from above and funnel concentrated beams to their algae gardens.[66] If a person built a garden at the bottom of a deep cavern and erected a series of mirrors to focus sunlight down to it, he'd be considered a great eccentric—but also a certain kind of genius.

Passage to the future

In the past 500 years, few lost causes were pursued more doggedly than the Northwest Passage: the fabled Arctic shipping lane between Europe and Asia. The islands and channels of northern Canada tempted explorers to navigate them in summer, only to watch their ships crack like eggshells in the grip of winter ice. The Arctic Circle's brutal conditions thwarted even the toughest expeditions. Time and again, nature crushed their hulls, froze their riggings, and turned them back.

The Arctic's majestic, bleak expanses are alien to our warm-blooded and terrestrial species. Plains of blinding white are sparsely populated by polar bears in feeble violet sunlight. Below the ice, through the proverbial look-ing glass, are tiny gaps and channels between ice formations. Wintering krill huddle in the safest crevices. Farther out to sea, belugas and narwhals glide through wider arteries—drifting like white ghosts, squealing sing-song at one another. And out in the great open spaces, bowhead whales gorge on roil-ing clouds of krill. Baleen plates sieve them from the freezing water, sucking them down into maws black and bottomless as the sea. When they encoun-ter pack ice, the bowheads' skulls are thickened and hardened into battering rams, rock-hard and capable of punching through 2 feet of solid ice.[67]

Explorers sought the Northwest Passage for the riches and fame they might earn. In 1906, Roald Amundsen was the first to complete such a jour-ney. Piloting a converted herring boat with a six-man crew, he crossed the Arctic from Greenland to Alaska. It was a 3-year labor of summer sailing and winter waiting, camping for months at a time on the solid ice outside the fro-zen boat. At the end, once they had arrived in Alaska, Amundsen skied 500 miles to the nearest telegram station to announce the news.[68]

The now-legendary explorer made history by using a ship with a shallow draft, the better to hug coastlines where the ice was thinnest. He chose a crew small and tough enough to live off the land for two harsh winters. They learned a great deal about the Canadian Arctic's native people and calculated the exact position of the Magnetic North Pole.[69] But ultimately, it was a hol-low victory. The Northwest Passage did not open up a new commercial ship-ping route or a vast new territory for exploration. The route was too slow, the channels too shallow, the ice too fast—and within a decade, the Panama Canal would render Amundsen's quest largely obsolete.

Much later in the twentieth century, biologists discovered the North-west Passage had been conquered many times before: by Arctic Ocean species crossing from the Pacific to the Atlantic. A huge multispecies exodus occurred 3 million years ago, in an event dubbed The Trans-Arctic Interchange. Hun-dreds of mollusk, fish, and echinoderm species invaded the Atlantic from the North Pacific, completing the journey over many generations.[70] Some of the Atlantic's best-known seafood, like cod and mussels, arrived during this period. The Northwest Passage has been mostly frozen during recurrent Ice Ages—even in summer—but it lies tantalizingly agape in the time between them.[71]

The biggest creature to traverse the Arctic may have been the gray whale, some 125,000–140,000 years ago.[72] Their ancestors lived only in the North Pacific, migrating to the Atlantic while the Northwest Passage stood open.

They were hunted completely out of the Atlantic in the eighteenth century, and that hemispheric extinction wiped out any DNA evidence of the great migration.[73] Recent work on whale bones from ancient Native American middens produced DNA from ancient Pacific gray and humpback whales.[74] Similar work on the Atlantic gray whale hasn't been published. We don't know for certain whether these whales traveled the Northwest Passage 100,000 years ago or 3 million.

With time, humanity took its revenge on the Northwest Passage, changing it from a lost cause to a predictably open sea lane. We didn't do this by developing better navigational tools or ice-breaking ships; we did it by melting the ice. Arctic ice melts as atmospheric CO_2 levels rise and as global warming advances.[75] The summer melt now opens the Passage to commercial shipping for weeks or months at a time.[76] It grows every year and will for the foreseeable future. It is a thousand-mile proof of the reality of climate change.

CHAPTER 10 THE STRANGEST FAMILY LIVES

*Between sex changers, egg biters,
gonad stealers, and more, the extreme
families of the sea succeed in many ways.*

Twists in a tail

Few ocean organisms have benefited from popular culture more than the merry clownfish. Just a few inches long, it's typified by bright colors and a friendly demeanor. Of the thirty-odd species in genus *Amphiprion,* the most popular aquarium types are neon orange with bold white markings. Those marks are broad and gently curving, guiding the observer's eye along the fish's rounded body. With clownfish, nature hosts a seminar on graphic design. Like the best consumer products, they're bright, distinctive, and eye-catching without drifting toward tacky. They are also famously loyal, spending their lives in symbiotic bonds with sea anemones. Specialized mucus protects them from the host's stinging tentacles. Flitting imperviously in and out of the poisonous thicket, clownfish clean their hosts and feed off the resulting detritus.[1] Predators won't risk a nasty sting for such a spare meal.

The 2003 Disney film *Finding Nemo* formally canonized the anemone dweller's adorability. The eponymous clownfish vanishes from his home anemone, forcing his widowed father to take off after him. *Finding Nemo* gets many things right—the anxiety of leaving home and the obnoxious yelping of seagulls—but it punts away the most fascinating aspect of clownfish. As *sequential hermaphrodites,* they lead unique home lives. All are born male, with the ability to change sex. Like a wild card, it's only good once: once males turn into females, they can't turn back into males. The film supposes a lifelong romance for Nemo's parents, but genuine clownfish live only as part of larger groups. A handful of fish share each anemone, all beginning their lives as immature males. The largest and most dominant male turns into a female; the next-largest develops functioning testes. She lays eggs, he fertilizes them.

The others bide their time, defending the anemone and the family's precious eggs.[2] One of the mated pair will eventually die, to be swiftly replaced by someone down the ladder.

If the matriarch dies, the fertile male who was #2 now takes her place as #1, metamorphosing into a female himself. A simple hierarchy of size and strength determines the family's whole structure, conflicting with the acceptable social norms for children's movies. *Finding Nemo* painted a simple picture for more than just the sake of simplicity: a real clownfish father who lost his mate would not develop a psychologically complex system of grieving and overprotection. He would simply become Nemo's new mother. Nemo (the only other fish remaining in the anemone) would rapidly develop mature gonads. He would become his own father while his father became his mother, and they would raise little incestuous Nemos together without a drip of sentimentality. In retrospect, the producers at Disney probably made the right call.

Anglerfish

Across species, males will endure great trials to find mates. They'll hurl themselves into combat, develop elaborate ritualized displays, or pay absurd prices for drinks at bars. But nobody gives up more for sex than the anglerfish males of the deep ocean. These animals give a new dedication to the phrase "mating for life."

Deep-sea bottom-dwelling anglerfish are among the sea's most hideous animals. Wrinkled black skin stretches over feeble muscles. Beady black eyes stare above a hatchet-like jaw, protruding tongue, and glittering needle teeth. The typical anglerfish sports a long jointed stalk (the esca) atop its head: a dorsal fin modified into a lure, waggled on command to simulate a floating morsel of food (see Chapter 4). Most deep-sea anglers store symbiotic bacteria in their lures, letting the fleshy knobs shine like beacons in the black. The fish hangs inert in the water column, waiting to strike with incredible speed. Touching the lure triggers the angler's bite reflex—insurance against a sloppy mistake, for it cannot afford to miss a meal in the starved ocean depths. Neither must it be picky; its jaw and stomach are extremely elastic, capable of devouring animals twice its size.[3]

For a century, marine biologists considered anglerfish rarities, seen dead on beaches or found in deep trawling nets. From those few samples, they noted two curious things: every specimen was female, and most adults car-

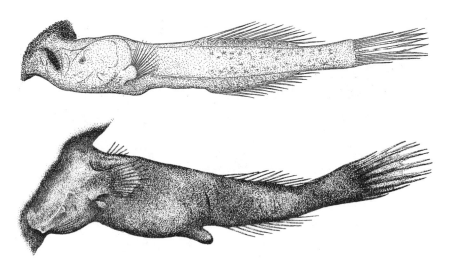

Parasitic male anglerfish (*Neoceratias spinifer*) showing structures that attach him to the female From Pietsch, Theodore W. 2005. "Dimorphism, parasitism, and sex revisited: Modes of reproduction among deep-sea ceratioid anglerfishes (Teleostei: Lophiiformes)." *Ichthyological Research* 52(3):207–236, figure 17. Courtesy of Zoological Museum, University of Copenhagen.

ried odd fleshy parasites attached to their bodies. These weren't remarkable findings for such a rare and mysterious fish. But in 1925, British ichthyologist Charles Regan took it upon himself to thoroughly dissect an anglerfish parasite.[4] He was shocked: it was an anglerfish male! Adult males never really existed to be found. They lived only as blind parasitic dwarves, attached permanently to much larger females.[5]

It's an extreme example of *sexual dimorphism*: inherent physical differences between sexes. Female anglers are vicious and successful hunters, able to devour most organisms they encounter. The great abyss holds little to threaten them. A male is the polar opposite, tiny and helpless. His undeveloped digestive tract means that even if he caught prey, he couldn't eat it. Between predation and starvation, he isn't long for the world.[6] But he has one thing going for him: some of the most sophisticated sensory organs in the deep ocean. Olfactory glands for some anglerfish, enormous light-gathering eyes for others—whatever the species, the male's senses are finely tuned to detect females.[7]

He's in a desperate race against time: find a mate and attach, or die. He soldiers on through cold black water guided only by hunger, instinct, and smell. When a female finally comes into view—a grotesque monster dozens of times his size—our hero acts immediately. He bites her, latching on with every

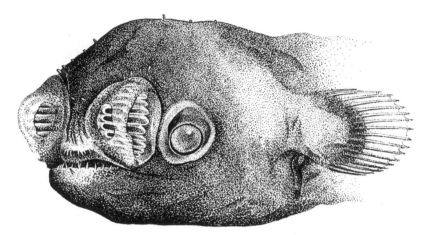

Free-living male anglerfish (*Linophryne arborifera*), showing huge nostrils and eyes to detect females. From Pietsch, Theodore W. 2005. "Dimorphism, parasitism, and sex revisited: Modes of reproduction among deep-sea ceratioid anglerfishes (Teleostei: Lophiiformes)." *Ichthyological Research* 52(3):207–236, figure 18. Courtesy of Zoological Museum, University of Copenhagen.

ounce of strength in his feeble jaws. This union releases an insidious enzyme in his own body, which will both save and end his life. His lips and mouth dissolve as he chews, liquefying and binding him to the female's flesh until the pair are utterly fused. He can never leave her.

This cheery scene is just the first stage of mating. In the following days and weeks, the parasitic male's circulatory system merges with his host's. Her blood delivers nutrients, pumped through a shared lattice of new vessels and capillaries to sustain him for as long as she lives. His old form diminishes with time. Fins aren't needed, nor his amazing eyes—nor, at last, his brain and internal organs. They dissolve until only one system remains: the testes. The female has little to do with this process, barely noticing his paltry calorie consumption and, on occasion, chemically inducing him to release sperm. He doesn't even get the consolation of monogamy! Numerous males will attach themselves to a female over her life, all suffering the same gruesome fate for a chance at reproduction.[8] A male angler begins his life as a tiny, pathetic dwarf in a darkened netherworld. If he is very lucky, he may end it as a pair of disembodied gonads.

Palolo worm

It's a late spring night in American Samoa. Waves crashing on the reef churn out white noise even at great distance. Cool breezes cut through sultry air, and sudden showers lash the steep forested hillsides in 90-second spurts. A pickup truck rattles by on the coastal road, packed like a clown car with boisterous passengers. Another truck passes; more follow. Soon the road is a stream of vehicles and blaring headlights. Samoans carry fine nets in their hands and excited smiles on their faces. Across the islands, text messages hum into salt-streaked cell phones: *the palolo worms are spawning.*

Palola viridis snake through shallow tropical reefs, hiding from predators in coral crevices and sandy burrows. They're bristle worms: segmented annelids with leg-like bristles, like pink stubby-legged centipedes about a half-foot long and the width of a spaghetti noodle. A unique and fascinating reproductive cycle sets the worms apart from their many reef cousins. On one or two nights per year—coordinated perfectly with the moon's phases—palolo worms surge off the reef to spawn all at once.[9]

The preparation begins a few weeks before. Sensing the mating season's approach, the worm begins a metamorphosis. Intestines dissolve, new muscles develop, and the gonads begin to grow at a massive rate. When at last spawning night arrives, the worm's back third is a series of bulbous sections like cars on a freight train. Each carries powerful swimming legs and a bulging load of gametes: eggs or sperm, according to the worm's sex. As spawning time approaches, palolo worms bury themselves head-down in the sand to wait.

Pale moonlight filters down through the water, marking time somewhere deep in the worms' ancestral memories. In October or November, when the moon rises near midnight and the tide is rising, the palolo worms suddenly break apart. Their rail-car tails are severed from the larger animal and, blind and brainless, they swim out in to the night. Propelled by their paddles and sniffing out the thin moonlight with primitive light-detecting eyespots, the segments (called epitokes) haul their genetic cargoes to the water's surface. There they meet their fellows: millions of epitokes all released in concert across the reef.[10] Their sides burst open, flooding the shallow water with eggs and sperm. In minutes, the lagoon is a soup: thick and clouded with goopy ropes of protein. The epitokes are remote drones, programmed with a simple list of instructions and cast to the sea's mercy with the blind trust that millions of other worms will do the same. And by joining in the swarm, each one is safer from the predators on the reef, which are overwhelmed by the mass of spawners.

To Pacific Islanders, palolo spawners are greater delicacies than caviar or even gravy-slathered beefsteak with fried eggs. Hundreds of people leave their workaday lives to moonlight as fishers, wading in waist-high water with their nets and buckets. Between mucousy scoops, they'll toss a few in their mouths like impish children at a berry farm. The headlights of countless pickup trucks glare from midnight to three in the morning: artificial moons to lure the worms closer ashore. Whether eaten raw by the slurpful, pulled directly from the sea, or fried with onions into a salty omelet, the next few days are an orgy of palolo feasting across Polynesia.[11] Gallons of epitokes are sent overseas in coolers to relatives, frozen for special events throughout the year. They are the Samoan equivalent of Christmas cookies.

Back on the reef, the worms huddle in their tubes and nurse their depleted bodies. They've lost a large portion of their weight and over the next year must regenerate their tails to spawn again.[12] Under cover of night, protected from predators, palolo worms keep chasing the moon.

Dads of the sea

Parenting is an unequal burden for most animals. On land, females usually haul the load. If they're lucky, a mate will hang around to guard territory or feed the chicks. Ocean creatures are more egalitarian: both sexes tend to ignore their offspring, cranking out eggs and sperm in unbelievable numbers and leaving them to the vagaries of fate. Whether eggs are fertilized later by males or eaten by predators is of little concern. Parents usually give no thought or time to parenting.

A handful of species, tucked away throughout the planet's oceans, go against the grain. It's often not much: moms may protect broods of fertilized eggs for a while before pushing hatched larvae, tiny and helpless, into the world. Or they might carry them for a time. But if these moms are rare, attentive fathers are even rarer.

Seahorses

Imagine a traditional Chinese herbalist's shop—cramped and cluttered, with the musty-yet-venerable feel of an old library. Boxes and jars are heaped in the corners one atop the other, the highest capped with layers of dust like Alpine summits. The most popular remedies have been herded to the center of the room in cardboard barrels lined with plastic trash bags. The store owner glowers over his wares like he suspects you might steal them.

In one prominently placed container wait tiny green things like intricately gnarled lima beans. As you pick them up, you notice those features are really a thin tail, knobbly skin, and a delicately fluted nose, all curled into a fetal position. It's a seahorse, caught and dried and sold by the million as herbal

remedies for a host of afflictions. The animal is one of the planet's only prac-
titioners of male pregnancy.[13]

Human beings have known about seahorses for centuries, but the creatures have always kept an air of mystery about them. Though genus *Hippocampus* looks like nothing else on the planet, these creatures evolved from
pipefish—whose tube-like mouths they've retained.[14] Poised upright in the
water with tails demurely curled, seahorses propel themselves with a single
small dorsal fin.[15] Pectoral fins sited behind the eyes allow the animal to steer.
They are pathetic swimmers, lacking tail fins and puttering between perches.
Even weak currents knock them about, so their prehensile tails reach instinc-
tively for anchoring vegetation. Found in shallow water across the globe but
particularly in tropical climes, seahorses rely on algae, corals and sea grasses
for protection. They are masters of camouflage, lying in wait to inhale small
crustaceans and other plankton. The seahorse is probably the least intimidat-
ing ambush predator on the planet.[16]

But it compensates for its awkwardness with a fascinating love life.[17] A
courting pair meets each day at dawn. They engage in a dance, with colorful
choreography unique to each species: bowing, swimming in unison, changing
skin color, even locking tails together in a delicate embrace. After a few minutes, they part, only to meet again the next morning. The ritual continues for
up to a week, as the female's body rapidly "cooks" her eggs into a fertile state.[18]

When the pair is ready, they get to the actual mating business: a denouement distinct from the dawn dances. The pair rises from the sea bed, tails
entwined like ivy ropes, rotating slowly in unison. Their trunks lock together
in a sort of kiss at the spire's apex. The male's egg sac—running down the
outside of his throat to his belly—flares open to receive the female's tube-
like ovipositor. She transfers hundreds of eggs in seconds, growing slender
while the male's abdomen balloons.[19] All the while, he secretes sperm into the
water near the sac's opening. Wily gametes instantly seek out the eggs.[20] His
body now pregnant with developing embryos, the male goes about his business. The female's material role is done, though she keeps visiting her mate.
Every day they repeat the pre-mating ritual as though the pregnancy never
happened.[21]

In a few weeks, he gives birth in grotesque fashion. His body engorged
with up to 1,500 offspring, he contracts the sac to expel them in rapid-fire
explosions. Wriggling baby seahorses burst into the world like flak over Nor-
mandy. Tiny fins beating, they rapidly disperse to fend for themselves.[22] The
male retreats, exhausted, his work done.

A mated pair may stick together for a whole breeding season, or a male might swim away looking for a different mate for the next brood.[23] Seahorses are vain and fickle. Some species prefer mates of matching size, whereas others favor large females for their superior egg production.[24] The closely related pipefish are the ultimate breeding opportunists. Males generally prefer the eggs of larger females, and a pregnant male pipefish may simply abort his current brood if a better female comes around.[25]

Sergeant major damselfish

Rearing the next generation is a tough job, and sometimes it takes a tough attitude. No fish in the ocean projects more parental ferocity than the aptly named sergeant major damselfish, in the genus *Abudefduf*. Some species live on Caribbean reefs, others in the tropical Pacific; all are energetic and brightly colored.[26] Five prominent vertical black stripes measure out their 6-inch bodies, and splashes of brilliance may be found on heads and tails. They eschew weapons, refusing to arm themselves with spines or needle teeth or burning venom.[27] Like many undersized scrappers, male damsels compensate with extreme and unchecked aggression. They stake out territory on the reef and defend it furiously against all comers.[28] Parrotfish the size of small dogs drift into the zone, only to flee under sudden attack. Curious divers wince at the damselfish's pugnacious nips.

Females lead calmer lives, grazing on algae and plankton while building egg hoards in swollen bellies. On either the new moon or the full moon (depending on the ocean), the intrepid ladies head out in search of mates.[29] The males are prepared, having cleaned their territories much the way young human men frantically scrub their bathrooms before dates. Their neighbors are the competition (talking about the fish again), and females will carefully inspect each territory. Using a complex and mysterious set of criteria, they weigh a serious decision: to whom to entrust their precious eggs?

When a female arrives, the male travels out to his plot's boundary for a formal greeting—displaying a cordiality he extends to nothing else. With colorful displays and rapid movements, he invites her into his home. She inspects him and the nest carefully: how many eggs are already there? How many of them are near to hatching? She's more likely to trust her eggs to an established, successful male.[30]

Should she accept, the male leads her to a sheltered patch of reef rock. He's already stripped it bare of algae, corals and everything else, and it's here she deposits up to 20,000 eggs. He jets them with sperm as she swims away, and it's done. The entire process happens in less than 20 minutes. The two will never meet again. Steely-eyed males are now committed to non-stop egg defense, though they'll take a moment to woo other passing ladies.

Once the eggs are laid and fertilized, parenting strategies vary widely. In some species, males tend their eggs like precious gardens, weeding out dead specimens so they don't contaminate others. Others simply squat and guard.[31] For all his attitude, the damselfish can be overwhelmed by schools of invaders, and may (if defeated) join them in devouring his progeny. In other species, he may simply grow peckish and snack off the eggs.[32] Once the speck-sized larvae hatch, they drift away into the water column to fend for themselves. Sergeant majors make dedicated if two-dimensional fathers.

Elephant seals

If you're ever planning a San Francisco vacation, go in January and rent a car. At the crack of dawn, pack yourself a sandwich and start driving south down the legendary California Highway 1. It snakes along the Pacific coast, winding tortuously along 500-foot cliffs that fall straight down to a grim gray sea. The road is long and slow, but it provides some of the most spectacular coastal scenery in the world: wind-worn sandstone bluffs, guano-stained promontories of granite, thick sleepy heads of cypress trees peering through fog. Between San Francisco and Santa Cruz lies a lost coast that has fought the twenty-first century to a standstill. Lichen-crusted houses and farms hidden back in gnarled, thicketed gullies create an atmosphere of funky hometown decay.

Along this coast, off an unremarkable exit from Highway 1, is a world completely divorced from the Bay Area's tech-oriented strivers. Here the day's drama focuses not on venture capital but on battles between multi-ton monsters. Just 20 miles north of Santa Cruz lies Año Nuevo State Reserve: breeding grounds for the mighty northern elephant seal (*Mirounga angustirostris*).

The largest pinnipeds in the ocean—including not just seals, but also sea lions and the walrus—elephant seals are named for their vast bulk and the flabby elongated nose sported by the bulls.[33] "Bull" is an appropriate word here. Male elephant seals stretch 16 feet long and swell to the width of a passenger van: a 2-ton monolith of muscle and blubber. Females are much smaller, weighing about a quarter of a ton. They more closely resemble "normal" seals, and do not have the male's territoriality. They are, however, aggressive in defense of their pups when massive males lumber obliviously across the sand.

Elephant seals spend the vast majority of their lives at sea, eating. Master swimmers with physiologies fine-tuned for deep cold dives, they plunge thousands of feet to scoop up bottom-dwelling fish and squid. They go months without seeing land, summering in the far northern Pacific, where ocean gyres funnel prey into a chilly cornucopia.[34] When elephant seals finally make their way ashore for the annual winter breeding season, thick stores of blubber tell gluttonous tales.

The largest bulls are the first to land, settling on the beaches where the females will eventually arrive to give birth. The males look like gigantic brown boulders until suddenly they move, so fast and agile it's unnerving. For all their bulk, males can rear up and *run* at terrifying speed with a lumbering legless gait like a grizzly bear running a sack race. It's funny to imagine, but few humans could outrun an elephant seal bull on sand.

Males are highly aggressive, focused entirely on battles for status. The strongest bulls have the easiest access to cows, so the objective is to claim as many cows as possible and then to keep competitors away. While the big bulls fight over dozens of cows at a time, younger and weaker males tussle along the periphery. They won't challenge the big dogs this year, but these skirmishes prepare them for later opportunities.[35]

Mothers are the last to arrive. A bright-eyed young female flops her way ashore, doubly swollen with fat and pregnancy. She must first decide which bull's territory she'll live in, picking her next pup's father and joining his harem of other females. The second order of business is giving birth, which she does in the clammy sand without much fuss.[36] The new mother turns to bond with her pup, meeting its kitten mews with soft soothing grunts of her own. She'll spend the next few weeks nursing, losing more than 40% of her weight in the process. Neither males nor females eat at all during the beachbound mating season.

In the very last days of weaning, her depleted body becomes fertile once again. She'll mate with her chosen bull; if he's dethroned, she'll mate with the winner. Eventually she heads back to sea carrying a slow-gestating embryo in her womb.[37] For nearly a year she'll feed voraciously, both rebounding from birth and nourishing her next pup.

Cows live in relative tranquility, but bulls must clear one of the sea's highest hurdles for fatherhood. As the shore fills early in December, a brutal tournament begins. Even before the females arrive, the males are squaring off for domination. Boundaries are crossed, challenges are issued and sparring turns bloody for those males strong enough to play the game.[38] A veteran's chest, padded with blubber and armored with thick calloused skin, resembles a medieval knight's shield. The seals use this bulwark to joust.[39] With a booming challenge—half belch and half roar—floppy-nosed cavaliers square up and charge.

Heads reared back, these rhinoceros-sized animals collide with a thunderclap. Grappling, striking at each others' throats with gleaming ivory fangs, they're quick to draw blood. In moments it flows freely down their cornified chests. But their savagery has limits. When at last one submits, he's permitted to slink away—combat is brutal, but grave injuries are rare. It's an unspoken, instinctive pact among warriors: none on the beach wants to die over dominance.

The victor bellows in triumph through his flabby snout. Its intricate matrix of cavities serves two ends: first, it recycles his breath to save moisture and ease the season's physical strain.[40] Second, the proboscis amplifies his already formidable voice to better display dominance. The most successful males generally have bigger proboscides, and the correlation makes this an auspicious trait for mating.[41] When the wind is still, the seals of Año Nuevo can be heard for miles.

If the alpha male claims a spot on the beach too soon, constant combat may exhaust him by the time his harem becomes fertile.[42] If a male lands too late, the best grounds will be occupied by fierce defenders. The stakes for these battles couldn't be higher: the vast majority of elephant seal males never get a chance to breed.

As Shakespeare once wrote, "Uneasy lies the head that wears the crown."[43] Dominance is a tremendous strain. A dominant bull may hold 50–100 females in his harem at any given time, watching them come and go with their breeding cycles. Alert to rivals that may swoop in while he rests, or wait in the water to ambush cows, the alpha male responds to all challenges. Bulls mature sexually at about 8 years, and the lucky ones can snag a few females right away, but even the greatest seldom last longer than 4 years atop the totem pole.[44] Many males die from exhaustion after a single breeding season as alpha. Still more will not survive the long year at sea.[45] Yet their sacrifice yields great rewards: annually, the top five males at Año Nuevo father 85% of the next year's pups.[46]

Devoted mothers

Nest guarding by fathers and male pregnancy in seahorses shouldn't give ocean mothers a bad reputation: females that care for young are generally more common in the sea. Crustaceans carry eggs in chitinous clasps beneath their abdomens for months at a time. Corals and sponges gestate their tiny larvae within their bodies until releasing them to swim away. But the most devoted marine mom might be the octopus, living entirely for her eggs and dying once they hatch.[47]

The giant octopus of the Pacific Northwest (*Enteroctopus dofleini*) is fertile at just 3 years of age—remarkable speed for an animal measuring 10 feet from arm to arm. First, a female accepts packets of male sperm called spermatophores, tucking them away before seeking out an ocean-bottom burrow. Then she carefully piles rocks around the entrance, slips in through the very last opening, and finally seals herself inside. The roofs and walls of this octopus's garden soon are decorated with ghostly white eggs. They're carefully tended, bathed in jets of high-oxygen water from the mother's muscular siphon. She'll stand guard for more than 6 months, starving herself until finally the eggs hatch. At that point, with her larval young teaming about her like dust motes, she hurls her last ounce of strength at her burrow's gate. It cracks—just a few stones displaced—and through this aperture pour thousands of quarter-inch babies.[48] The mother watches the next generation depart, sinks back into her hole and lets the cold creep in. Octopuses don't so much live in their bodies as pilot them, the higher cognitive centers detached from fine motor control, so she might not experience death as suffering.[49] It may be like a grand theater at evening's end: a slow extinguishing of lights.

Most octopuses live fast and die young.[50] Females die nobly in the line of generational duty, but male giant Pacific octopuses suffer an insidious encroaching dementia. Eventually, this mental senescence leaves them swimming in such aimlessly frenetic patterns that they become easy prey. So tightly regulated is octopus senescence that it can famously be halted by brain surgery.[51] Removing the optic gland averts early death but also leaves the animal blindly unable to mate. Most octopus species live less than a year—some die after 3 months; snuffed out after their single breeding season.[52] The contrast to long-lived fish is extreme, and it's remarkable that such highly advanced animals crawl into such early graves.

Sea squirts—Divide and conquer

Botryllus schlosseri is a common variety of sea squirt. Properly dubbed tunicates, these animals appear on rocks and pier pilings as thin films of brightly colored flesh. Comprised of countless water-pumping units called zooids

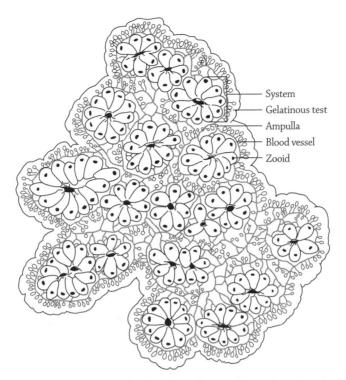

System
Gelatinous test
Ampulla
Blood vessel
Zooid

A colony of the tunicate *Botryllus schlosseri* ready for the gonad wars. This colony shows nearly 20 groups of zooids, called systems, each with a halo of ampullae and blood vessels around it. The ampullae and blood vessels are central to attack and defense strategies that ensue when two colonies meet. The spoils of this war for space are the ovaries and testes of the zooids, which may be taken over by stealth invasion of cells from adjacent colonies. Originally published as figure 1 in Milkman, Roger. 1967. *Biological Bulletin* 132:229–243. Reprinted with permission from the Marine Biological Laboratory, Woods Hole, MA.

(pronounced ZO-ids), each colony spreads into multicolored sheets. A single tunicate—a tiny zooid—has vital organs that are loosely bound into a gelatinous core and surrounded by cloudy mucus. A casual observer might take tunicates for anemone-like primitives; in truth, the sea squirt is a highly sophisticated creature. It's a chordate; a member of the same phylum as human beings (along with fish, birds, mammals, and countless other animals). Tunicates lack bones, eyes, a brain, or most traits we associate with advanced evolution—but hidden inside are the primitive beginnings of a spine. Full digestive and circulatory systems round out the squirt's complex anatomy, along with both male and female sex organs. Every zooid in a given colony is a clone of every other, connected through a shared blood supply.[53]

Botryllus colonies grow in cold, highly productive water in patches of bright colors—red, blue, yellow. They grow quickly during summer months, especially on docks and pilings in temperate harbors, such as Monterey Bay or Roscoff, France. When two colonies run up against each other, physically impeding each others' growth, they jostle to circumvent the restriction. They scout into enemy territory, each colony probing the other's edge with finger-like ampullae until their two bloodstreams actually mix.[54] The mixed individual is called a blood chimera, which sounds like a fearsome opponent from a video game. If the intertwined colonies aren't close genetic relatives—if their proteins can't cooperate—then a powerful reaction rips through the connected ampullae: blood congeals and clots, tissue grows inflamed, and the colonies are forced apart.[55] The parties go their separate ways—attempting to expand in different directions, away from the conflict zone—and interact no further. This inflammatory reaction isn't too common; new *Botryllus* colonies tend to settle in close proximity to their relatives, with which they tend to be more compatible.[56]

But if the colonies are compatible, a different sort of battle commences. The blood chimera becomes a true chimera (presumably with more hit points): a mixture of different individuals coexisting in the same colony. They merge together, freely exchanging blood cells and nutrients. But this cooperation is a façade. All the while, the colonies secretly battle for supremacy.[57] Releasing phagocytes and other immune cells into the shared bloodstream, one colony samples the other's body chemistry. A cold war has begun, a contentious relationship with both sides engaged in acts of biological espionage. Specialized cells infiltrate an enemy zooid, seeping into it like special forces troops capturing an enemy building. The infiltrating colony then rebuilds the target zooid from the invading colony's tissues. From that moment on, that zooid belongs to the invaders.

The process goes both ways, each colony annexing parts of the other. Like bitter feudal families, they war back and forth and ultimately form one genetically heterogeneous whole, like a patchwork of villages held by different warlords. It is one solid crust on the rock surface, studded with zooids that have different genetics and are made from the genomes of different individuals. The colony may be able to use its enhanced genetic diversity to adapt to a changing environment, or it may conquer the next colony it encounters and claim more territory.[58] Either way, total victory is rare, and the most common result is a painter's palette of differently colored zooids across the substrate.

But a still deeper conflict rages all the while, nearly hidden. Some cells from the immune system are even more insidious than their special forces comrades. Rather than targeting whole zooids, they target only their gonads to be rebuilt from the invader's tissue. The rest of the zooid keeps its original cells and genes. From that point forward, the hapless zooid keeps feeding its gonadal tissues, but those tissues will produce gametes for the invaders instead. The war of the gonads plays out silently for sea squirts, one zooid at a time, until the whole of the invaded colony has the usurpers' gonads. The invaded colony is now living a lie, like a royal bloodline polluted by a common stableboy. Instead of true offspring, the fruit of its zooid loins are foster children carrying the usurpers' genes. For as long as it lives, this subtly conquered colony will produce only the conqueror's offspring.[59]

Happy diversity

Leo Tolstoy once wrote that happy families are all alike, but "every unhappy family is unhappy in its own way."[60] This might hold true for human families, but in the sea it disintegrates like soft bread. Beneath the waves, even happy families hold nothing in common. What works for one may not work for another. The ocean's most exotic inhabitants don't share our common ancestry, or our highly specialized sex organs, or even our concept of gender. Nature cares for the ends of reproduction, not the means.

Is it better to live fast, reproduce quickly, and then die, like the octopus? Or would you rather live a long time, investing ever more in every egg as you age? Steve Berkeley of the Long Marine Lab in Santa Cruz, California, has exhaustively charted the life cycles of many fish. A kelp forest species called the black rockfish (*Sebastes melanops*) starts producing eggs at 5 years old.[61] Even at that tender age, they're laying 45,000 eggs at a time. But unlike octopuses, these fish enjoy many annual breeding seasons. At 10 years of age, they've grown to 16 inches and may produce 60,000 eggs. Some closely related rockfish live up to a century (see Chapter 6), growing all the while and manufacturing more eggs with each passing season. For these fish, long life and future reproduction is the strategy. This differs wildly from the octopus plan.

All this shows that there is no single answer to whether long life or fast reproduction is better. Well, there *is* a single answer but the answer is "It depends." It depends on whether growing a body that can last 100 years is likely to pay off (in numbers of offspring) in the long run—or whether a cheaper, more disposable body lets you funnel more food energy immediately into reproduction early in life. There is no correct answer to this conundrum—it depends on how quickly an individual is likely to die from predator attack, starvation, bad weather, or other lethal events. High rates of death make it less valuable to invest in a robust body. But of course, a robust body makes the death rate drop. Does it drop enough to pay for the extra costs in body construction? Because this answer might vary for different kinds of bodies—fish versus squid, for example—the best evolutionary strategy also might vary.[62] The rules of the game are still the same: each species bets on either long lives or short, tailoring its strategy to the hand it's dealt.

Although this huge variation in how marine species live their lives is wholly understandable in terms of normal evolutionary science, marine mothers do more than a few things that continue to confound evolutionary biologists. In the black rockfish studied by Steve Berkeley, older mothers give every egg an extra gift—a tiny oil droplet that the larvae use as an energy supply to grow faster and be more likely to survive.[63] The droplet is like a trust fund that helps a larva through a very tough part of a fish's life, when perhaps fewer than 1 in 10,000 will survive.

According to other fish biologists, the problem with this set of results, and other examples where bigger females produce bigger eggs, is that it just shouldn't happen.[64] It's not that larger mothers shouldn't make nice eggs and give their larvae oil droplet gifts. It's that smaller mothers should do the same. If a certain size egg is best for a large mother to make, it should also be best for a small mother to make. Now, maybe small mothers cannot afford such gifts to all their larvae—but in that case, theory says that they should make fewer larvae and go ahead and give them those same gifts. But fish like the black rockfish apparently do not read much theory. In this case, the ways of mothers remain something of a mystery.

What about some of the other strangest family lives? Can they be reconciled with our understanding of evolution's relentless pressure for reproductive success? Sex change is one of the most interesting questions: some species that change sex start off male and change to female. Some do the reverse. Some flip back and forth, or are both sexes simultaneously.

One of the interesting ways to think about this is to consider the level of parental investment in each offspring. For many marine species, eggs are cast out into the ocean to fend for themselves—no parental care is provided. The investment in an offspring is entirely accounted for by the food energy packed into it, like sending the kid off with one packed lunch for the rest of their lives. And the vast majority of that investment is made by mothers, not fathers, because eggs are expensive to make, and sperm are cheap. A small female can only make a few eggs, but a small male can produce many sperm to fertilize many eggs. So, for a species that reproduces early, at a small size, and then continues to grow, it can be best to be a male first—fertilizing lots of eggs—and then as he grows, to become a female able to make lots of eggs. Clownfish play precisely this game, and so do some limpets and shrimp.[65]

Why do the reverse strategies exist: starting out as small females and then turning into larger males? Caribbean blueheaded wrasses follow this strategy, and so do California sheepheads and many groupers. The answer is often territoriality: males are successful when they hold territories and socially dominate other males. Bigger males do this better than smaller males can, so being a small female turning into a bigger male is more successful than the reverse. Blueheaded wrasses, for example, start out as yellow striped females in a harem dominated by a large, territorial male. If the previous male is removed (by a predator or an inquisitive biologist), the largest female turns into a male, starting the very next day. Once the transformation begins, she starts acting like a male in a day, and starts making sperm within the week. Thereafter, he spawns every day with his harem, sometimes running out of sperm by late afternoon.[66]

Why can't big females hold territories and attract smaller males? They probably can, and do in some species. The point is not that one particular strategy is best. It is that different strategies can work well and appear to be beneficial in different settings. Figuring out what those settings are has allowed biologists to put some of the wide variety of reproductive life in perspective.[67] But it would be a mistake to think that we know all about the sea's oddest families, or why they persist where they do.

CHAPTER 11   FUTURE EXTREMES

*The future oceans—and what lives
in them—are ours to choose.*

Across the wide expanse of the oceans and their deep history, marine life is phenomenally tough. Marine species are the ultimate survivors: from microbes that conquered early Earth, to living fossils crawling from the Burgess Shale all the way to the Ocean City boardwalk. Some cling to boiling vents as the only food source in crushing depth and darkness. Others hew an existence from the bones of dead whales. Helpless anglerfish males comb the night for brides sustained only by faith and a gold nugget of yolk. Reef corals—just our current kind, never mind that there were others long ago—have survived five major extinctions and a quarter billion years of planetary perturbations.

Yet ocean life is also often fragile, living in highly specialized ways in highly specialized habitats. The secret of the extreme life of the sea is that it is extremely good at thriving in these difficult environments. And these successful species are often wonderfully adapted to a special niche—a particular way of life. On land such niches also occur, but they are often small—little fish living in skinny streams—so that they can only accommodate a few individuals of a specialized species. The oceans, however, are so big that even a relatively tiny niche can sprout whole ecosystems. Hydrothermal vents are a good example: they occupy a tiny, tiny part of the sea floor, and they each last only a few years. But the sea floor is so big that there are a lot of hydrothermal vents, and marine life has adapted to skipping between them as the vents open and close.

Icefish survive temperatures that freeze blood. Swordfish employ eyeball heaters to accelerate their vision and catch prey. These intricate adaptations to ocean life rely on the ocean being so big that even a niche survival strategy can work. Once we see them this way, extreme marine species are like boutique businesses—extremely good at what they do and dependent on a thriving economy.

Unfortunately, these specialized successes are pressured from all sides by human activity. Every day we make life harder in the ocean. It doesn't take much: a little more heat, a little more acidity, a hitch in the food supply, or an uptick in fishing. These changes have grown as the human economy has grown, and today the impact of more than 7 billion people presses down on the oceans. Humans affect ocean life all the way from the level of individual organisms to the functioning, and survival, of entire ecosystems.

What a difference a degree makes

If you had one foot in the coldest ocean habitat and one in the hottest, you'd straddle both the freezing polar sea and a seething hydrothermal stew. One foot would be at 30° F (–1° C) and the other at 250° F (121° C).[1] Both places support a bright mélange of life, from icefish to Pompeii worms: life that thrives in its specialized niche.

In contrast to the wide temperature range of ocean life, global climate change projections suggest an increase in ocean temperature of 4–5.5° F 2–3° C).[2] This may seem like a small number to us, because we are remarkably indifferent to climate. As big mammals living in air, with efficient heating and cooling mechanisms, we've got a warped perception of changes in heat. Our ancestors spread from tropical Africa to the edges of northern European glaciers in just thousands of years, warmed only by animal skins and open fires. Later they crossed the Bering Sea, spread to Texas, ambled across the Panama isthmus, took up residence in the Andes, and moved to the southern tip of South America.[3] Along this path, we woke up to many different climates, but our big mammalian bodies helped us cope.

A human body uses about the same energy as a 100 watt light bulb to keep us warm or cool.[4] When our body temperature drops, our metabolism speeds up a little to compensate.[5] Calories burn faster and muscle activity generates more heat. And of course when outside temperatures rise, we sweat to dump the extra heat. Throughout these processes, the rest of our bodies operate at roughly normal temperature and speed. Our digestion, thinking, heart rate, and other metabolic functions are only slightly affected by exterior temperature.

But for cold-blooded creatures, environmental temperature is everything. An increase of about 22° F (10° C) in external temperature causes their metabolic rates to roughly double. The clock of life ticks twice as fast, more food is needed, and more waste products are produced. For perspective, a human heart doubles its rate after a single flight of stairs.[6] Think of life at higher temperature as a perpetual workout.

This extra labor is difficult to sustain. San Francisco State University's Jonathan Stillman tracked the heart rates of crabs in slowly heated water. As temperature increased from 61° to 93° F (16° to 34° C), heart rate climbed from 148 to 403 beats per minute—and finally crashed as the animals suffered fatal seizures.[7] Stillman, no sadist, was investigating how close to this lethal temperature a typical crab lives. Surprisingly, he found that many crab species live pretty close to their temperature limits, inhabiting places where global warming's projected 4–5.5° F (2–3° C) increase would prove fatal. For them, the projected temperature change over the next 100 years would not just bump up metabolism, it would push these crabs over a heart-rate tipping point on the hottest days of the year.

Climate change is poised to do just this.[8] Already ocean species are on the move, shifting to higher latitudes, just as terrestrial species have.[9] But they also face a second man-made threat in the form of a corrosive acid.

Hot and sour soup

Human beings pump 9.9 billion tons of carbon as CO_2 gas into the atmosphere every year.[10] Nearly 25% of that enormous amount dissolves into the ocean, where simple chemistry takes over and the CO_2 becomes carbonic acid—the same chemical that lets children dissolve iron nails in glasses of soda.[11]

Across the planet, carbonic acid and ocean acidity[12] have steadily risen for decades. Acidity in the world's oceans increased 22% since the beginning of the twentieth century and shows no sign of dropping.[13] The risk isn't that marine creatures will melt into goo like the Wicked Witch of the West—they will suffer in subtler, but equally dramatic, ways.

Marine animals make their shells out of seawater, conjuring hard skeletons from clear fluid in an almost magic fashion. Accumulating calcium carbonate in tiny pockets of seawater surrounded by tissue, shelled organisms force the substance into crystals by maintaining a precise chemical imbalance. In this way, infinitesimal layers of new shell form along the edge of an

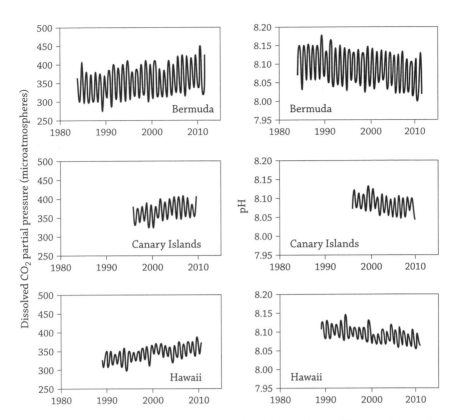

Increase in CO$_2$ (left) and decrease in pH (right) at three places in the world's oceans. Increased acidity results in decreased pH, so pH drops as CO$_2$ increases.

existing shell. A coral polyp sits on top of its own skeleton, built layer by layer over hundreds of years.

But the chemical imbalance needed to make shell requires low acidity levels, much lower than normal seawater. It's costly to reduce acidity for the purpose of shell-making, and the cost rises if the sea's more acidic than normal. Think of boiling water on a stove: starting with cold water takes more energy and time to boil than does starting with warm water. And in the ocean, making shell starting with more acidic water takes more energy and time for the shell-making creature.

Rich Palmer, a marine biologist at the University of Alberta, estimated that a typical snail spends more food energy on its shell than on tissue growth or reproduction.[14] Similarly, corals use 20–30% of their daily energy supply for skeletal construction.[15] Increased acidity makes the process much harder, and over the next century the price of hard components (both shells and bones) may rise 30–50%.[16] So, just as for high heat, acidity is a kind of metabolic tax.[17] At a certain rate of taxation, the cost is too high, the metabolic bank is broken, and an organism cannot survive.

In a more acidic future, many shell-building species are likely to suffer. Hundreds of experiments have tested CO_2 resistance in marine animals. Most show negative effects, especially for shell-builders, though the result isn't universal.[18] By the end of the current century, the predicted build up of heat and acidity will make it much more difficult for many species to thrive.[19]

Some of those species directly impact human welfare: oyster farms across the U.S. West Coast face reduced hauls stemming from ocean acidification. Like a young college graduate, the fate of an oyster depends sensitively on the climate in which it was spawned. For humans, entering the work force during a recession can cripple earnings for a whole career.[20] Likewise, oysters spawned into acidic seas face clouded long-term prospects. Batches of eggs yield fewer viable larvae when released into water with a high CO_2 and higher acidity. What's more, those larvae don't grow as quickly—their present weakness can persist, like human poverty, into the next generation.[21]

No new taxes

Ocean warming and acidification harm species in very different ways, but ultimately, their damage can be distilled to hard metabolic currency: calories flowing in from food, and calories burned by metabolism, growth, and reproduction. Acidification and overheating are, in their own ways, taxes on organisms' metabolic income. Every calorie spent on higher metabolism to deal with heat or acidification is one not spent on growth or reproduction. The tax rate may be low today (though some species are already pinched), but it climbs with every year. By the end of this century, it will have become truly onerous for many species across much of the world.

Of course, the most successful species—the ones with the highest incomes—can most easily pay the tax. Some reef-building corals can ramp up their feeding and stave off acid's burn.[22] Having the right stress-resistant genes allows a coral to live in warmer water or a sea urchin to grow in low pH.[23] Yet even these lucky species pay a price; every year of extra heat and acidity erodes their metabolic reserves. In the end, even the richest species will feel the metabolic tax of climate change.

More ocean, and the waves of higher seas

Warm water expands a little bit, and when a lot of water expands a little bit, the ocean rises. This kind of thermal expansion is thought to be the major cause of the sea level rise, with important contributions from glacier melt and other human-made causes.[24] As the ocean continues to warm, it will continue to expand, glaciers will continue to melt, and sea level will continue to rise. Even without the contribution from a catastrophic collapse of ice sheets into the sea, which could bump up sea level very quickly, the future of the ocean is that there will be more of it.

Average sea levels have risen 7 inches in the past 50 years, and the best science predicts a rise of 2.5–7.0 feet over the next 100 years.[25] For low-lying coral atolls, such as Funafuti in the atoll nation of Tuvalu, in the central equatorial Pacific, such a rise will nearly inundate the country—the highest point of Funafuti is 9 feet above sea level. The rise in the western Pacific and Indian Ocean is likely to be even greater.[26] This rising tide floods all shores, not just coral reefs. Salt marshes, like the ones that protect much of the U.S. east coast from wave damage, grow upward by only a quarter to a half inch a year and will be flooded if sea level rises as currently projected.[27]

Flooding is the most important consequence of sea level rise, not just because it damages habitats and alters the way many coastal species will live, but also because it promises huge damage to our human economy. Ten percent of all humans live in coastal zones, and that fraction is expected to rise.[28] Increased tidal height and storms have already created urban devastation in New York City, costing $60 billion to repair. New Yorkers are proposing to spend $20 billion more to try to guard the city's core against future catastrophes without even being certain that the proposed changes will work.

Around the world, the economics is similar—the cost of rising tides is the cost of coastal devastation by waves and storms, devastation that is often guarded against—now—by living marine life. Coral reefs catch the waves offshore and buffer erosion on the coast. Living corals absorb more wave energy than a dead, smooth pavement of former reef.[29] But when the tide and waves are high, or the corals are dead, the usually placid shores of a tropical lagoon can become a cauldron of churned up land and sand. Mangroves protect other tropical shores. Places that had intact mangrove forests showed less damage from the 2004 tsunami in Southeast Asia than did areas where communities had cut down their mangroves.[30] Coastal marshes, oyster reefs, and sea-grass beds play this kind of hidden buffering role on temperate shorelines.

Seawalls built to hold out the ocean might seem like a good solution to this problem, but they are expensive: $2 million dollars a mile in some cases.[31] But suppose there was a seawall that would grow itself, and would also grow as sea level increased, so that it provided continuous protection for the coast? A healthy, fast-growing coral reef or marsh can do that at about 3–4 feet a century. And it will provide this service for free across hundreds or thousands of miles of coast. All that these free seawalls need is for their ecosystems to be kept healthy and for the coastal zones to be managed so that they can grow.

Maintaining those healthy ecosystems in a changing climate presents some major challenges. But adding to those challenges is a new layer of ocean problems—when the species in the sea, and the way they interact with one another, are fundamentally altered.

Breaking the ecology of the sea

Once you understand a bit about the amazing life of the sea—from the smallest microbes to the largest whales—you can better see its incredibly interconnected nature. As in all closed systems, even gigantic ones like the world's oceans, several effects can combine into a feedback loop and unleash even bigger consequences. Oceans will not only be warmer and more acidic; their ecosystems will also operate under fundamentally different rules.

Marine environments are powerful biological machines, capable of staggering levels of productivity. Human fishing fleets remove about 88 million tons of fish and shellfish annually, which sounds like a big number until you realize the ocean's population of marine microbes produce this amount of biomass in about *an hour* (see Chapter 3). Left to pile up, this bounty would swiftly bury us all in sludge. It doesn't—because equally massive systems exist to consume that biomass. The burgeoning biomass is held in check by the normal ecology of the sea.

Microbes and single-celled algae fuel a worldwide food web that eventually produces sardines, whales, turtles, tunas, sharks, and every other extant marine creature. Human beings won't destroy that productivity—microbial life is so old and powerful, we simply can't—but for the first time, we can direct it. Humans are so numerous, so technologically advanced, and so widely polluting that our species can finally alter the planet's biggest habitat and its most abundant citizens, the microbes.

The effects pile up, quickly and literally: an unebbing tide of bacteria, algae, and other weeds all freed from the ocean's normal checks and balances. That unleashed and uncontrolled productivity is an ironic bloom of living things, but it wrecks the way the ocean naturally channels energy from species to species in the great chain of life. The combination is a fearsome feedback loop we call a Productivity Bomb. It has two primary fuses: one involving the most advanced marine life, the other the least advanced.

The first fuse: The war on fish

Decades of relative peace and prosperity following World War II led to an odd and undeclared global conflict: the War on Fish, where the expanding global population devoured everything the oceans could provide.[32] Humans won the war, naturally, but our victory was somewhat Pyrrhic. Rampant overfishing has reduced populations of fish across the world's oceans, from the tropics to the polar seas. We are simply taking more than the ocean can replenish—something we never realized was possible. This shift is unique in the planet's natural history. Never before could a single predator have such a massive global impact.

In the Philippines, off an island called Bohol known for its rolling "Chocolate Hills," sits an unusual coral formation: double barrier reefs. On a calm evening near the edge of the inner reef, Steve Palumbi gets ready for a night dive. The dive is led by Amanda Vincent of Project Seahorse, and seahorses are precisely what they're hunting. Both are dressed in wetsuits with tanks and regulators. Masks, flashlights, steel knives, watches, and depth meters round out their expensive equipment. The Filipino boatman, piloting his tiny *banca*, wears a fraying pair of swimming trunks along with goggles carved from a coconut husk and two disks of Coke-bottle glass. A kerosene lantern casts wan light from the bow. With soft splashes, each of the three divers goes over the side. Amanda leads Steve off to seek her favorite seahorses. The boatman takes off alone, frog-kicking down toward the reef, his bare feet a stark contrast to Steve and Amanda's flippers. He's got a family to feed and his own seahorses to catch—for sale rather than for science.

It takes Amanda and Steve 30 minutes to find just one seahorse, 45 feet below the boat. Elegant and enigmatic, it grips a coral branch with a delicate prehensile tail. It eyes the divers warily. A larger figure looms out of the dark water, and Steve swings his flashlight around to reflect off coke-bottle goggles. The boatman has doubled back in his hunt. With a wave he turns away, leaving them and the rare little animal in peace.

The scientists surface after an hour, cold and tired, unable to find a second specimen. The pilot ferries them back to shore, but once their gear is unloaded, he heads back out to sea. For the next 8 hours, he'll swim shivering through the dark water, diving again and again, finally returning near dawn with his own seahorse. It's only one animal, the length of your long finger, but he's able to sell it for 25 cents. It's enough to buy a few cups of rice. His wife and children will eat for another day. When the Sun goes down again, he'll return to hunt on the reef.

There were once plenty of seahorses in these waters; the men who fished them led easier lives. Only a few hours of work would yield a bagful of wriggling little flute-nosed fish. But as generations of fishermen took freely from the reef and the Philippine population exploded, seahorses became truly rare on the reefs.

Better hooks, fewer fish

Tales of local overfishing tug at the heartstrings because of the immediate impact on fishermen and their families. In the past, the fishing industry would respond with technology. For thousands of years, people fished from wooden craft propelled by wind or their own oars. They cast hemp lines and hurled harpoons. Modern fishing boats are battle tanks compared to these chariots. Satellite navigation, high-precision sonar, diesel winches, and deep freezers make these vessels technological marvels. If there are fish to be had anywhere on the globe, industrial fishing fleets will find and take them. But for all their sophistication, these high-tech armadas often don't take more fish than the simpler fleets of the past. How could this be?

Ruth Thurstan and Callum Roberts chronicled the history of one fishery in Scotland: the Firth of Clyde, a region of deep submerged canyons in the Irish Sea. Fed by the North Atlantic's nutrient-rich waters, it consistently yielded big hauls of herring, cod, and haddock through the early nineteenth century. As time wore on, bottom-trawling ships with steam engines replaced the old wooden craft. Bit by bit the fleet was modernized. Scouring the Firth's deep chasms improved fishing yields, but only temporarily. Thurstan and Roberts found that before long, despite these better ships, the catches dipped far below their previous levels.[33]

Callum Roberts cast a wider net with his research and found the same patterns repeated throughout Great Britain. Each time the trawling fleets took a technological leap, the catch would rise but then rapidly decline. Today in British waters, for all our technological achievements, it's twenty times harder to catch a fish than it was 200 years ago.[34] The fish are no smarter or more evasive—they're just less common.

And when a previously common fish is stripped from the sea, there are consequences for other species. For their predators, because food will now be scarce. And for their prey, because there are now fewer checks on the growth of the prey population. The result is an ocean ecosystem out of balance.

The great chain of life

The fish-eating public prefers to eat oceanic predators. We have decided they taste better: tuna, salmon, marlin, grouper, rockfish, orange roughy, halibut, and flounder all are fearsome predators in their natural environments. And when we eat these fish, we change the ecology of the oceans, because we are lopping off the top of the food chain. Daniel Pauly and his colleagues called this fishing down the food web.[35] Like blowing the tops off mountains in search of coal, this approach does massive collateral damage.

No one appreciates the unintended consequences of overfishing like scallop fishers from the Carolina coast. For generations they harvested bay scallops (*Argopecten irradians*), scooping millions of pounds from muddy seagrass beds.[36] As the twentieth century wore on, the scallop numbers started to fall. Down and down they slid, finally hitting rock bottom in 1994, when the entire season's harvest netted just 150 pounds.[37] The fishery collapsed, but not because of overharvesting the scallops themselves. In this case, the culprit was a salty soup brewed halfway around the world.

Shark-fin soup is a traditional Chinese delicacy that signals status and prestige. As China's economy grew over the past few decades, many millions of people suddenly could afford to serve shark-fin soup at their children's weddings and other important events. To feed this fad, the world-wide hunt for shark fins exploded over the past decades, and eventually came to the Carolina coast.

Most sharks are not really very extreme—their most extreme features are their teeth and evolutionary past (see Chapter 2)—but they are prodigious eaters. By the time they're full grown, they've consumed countless times their weight in lesser seafood. Off the placid Carolina coast, sharks make regular meals of cow-nosed rays (*Rhinoptera bonasus*), chomping them like leathery pancakes—that is, they used to. With the shark population depleted by international fishing fleets, the gentle cow-nosed rays were freed from most

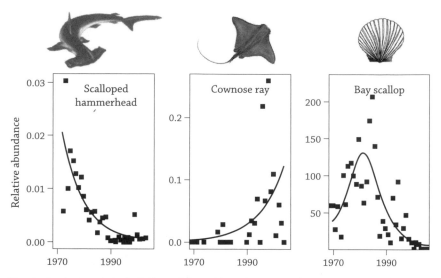

The decline in coastal sharks (left) was followed by an increase in their prey, cow-nosed rays (middle), and a steep decline in bay scallops (right). Redrawn from Myers, R. A., J. Baum, T. Shepherd, S. Powers, and C. Peterson. 2007. "Cascading effects of the loss of apex predatory sharks from a coastal ocean." *Science* 315:1846–1850.

predation. The rays frolicked through this new undersea utopia, reproducing in vast numbers and merrily devouring their own favorite prey: bay scallops. Thick shells were ground to broken china in the rays' flat powerful jaws, and one scallop bed after another was depleted. Fishermen found themselves desperately competing against the rays.

Here is the great chain of life at work: sharks eat rays, rays eat scallops, scallops feed us. Ecologists call this a *trophic cascade:* a natural chain reaction of collapsing and exploding populations. But when the cascade is altered, it can leave the sea fundamentally unbalanced. When the food chain is frayed or fractured, life accumulates just below the fissure. Kill the sharks, and their prey run amok. The backup of life at the break in a food chain is like cars backed up at an accident: the flow of traffic comes to a halt. But the delivery of cars along the highway doesn't stop, and so a traffic jam ensues.

The second fuse: Massive microbes

Imagine sitting in that traffic jam, because of an accident up ahead. As 5 o'clock comes around, thousands of people get off work and hit the roads.

The extra cars make the extant clog much worse. The same thing can happen in a food chain if discarded fertilizer kicks its smallest species into overdrive.

The Mississippi River drains 41% of the central United States, including the country's agricultural breadbasket.[38] Mountains of nitrogen, phosphorous, and potassium are slathered on American fields every year.[39] At least 3 billion pounds of discarded fertilizer wash down the Mississippi each season, most of the product flowing into the Gulf of Mexico.[40] That fertilizer is great for plants when it is on the farm fields, but it's also a microbial bonanza. The annual runoff creates ideal conditions for the single-celled algae that live in the warm, calm waters near the Mississippi's mouth. In days they can reproduce to make enormous blooms, forming dense surface slicks that color the water for miles.

The pattern is global. In 2010, runoff from French pig farms and other agriculture produced a bolus of green algae off the Brittany coast. Countless tons of thick green filaments washed onto the beaches and rotted in the Sun. The decaying piles were so foul that they incapacitated a horse and rider on the beach, killing the horse and necessitating rescue for the rider.[41] For the past decade these algae have regularly plagued the shore—the pig farms have never ceased operation—and more than a few animals have succumbed to the fumes.

Red tides, brown tides, and yellow tides have grown alarmingly common across the globe—all named for the colored algae that produce them and churn out noxious toxins. Shellfish and other grazers eat the algae, collecting the poison in their own bodies and potentially passing it to human beings.[42] In 2008 alone, more than 400 harmful algal blooms were recorded worldwide.[43]

Traffic jams and dead zones

A heavy tide of nutrients doesn't have to be a bad thing. In Antarctica, nutrient-rich upwellings from the deep sea feed seasonal megablooms of algae. Krill and copepods teem where the algae bloom, reproducing at planktonic rates and powering an ecosystem that's come to rely on such bounty. In March 2012, researchers from the Australian government's Antarctic Division noticed a floating bloom of algae 120 miles long.[44] Likely made from the single-celled species *Phaeocystis,* the bloom bound up billions of cells in a gooey mucus matrix.[45] Algae are a natural part of the ecosystem, and the polar food web consumed the bloom in mere weeks. In this case, the productivity explosion was blunted by transferring it from the bottom of the food chain to every other level, dispersing the excess energy and keeping the system balanced. In the absence of human intervention to sever the food chain or fuel further algal growth, there was no traffic jam.

The Gulf of Mexico has no comparable system left intact. Algae grazers like shrimp and oysters are depleted by overfishing, so the Gulf struggles with fertilizer runoff. The extra productivity takes too long to filter up the food chain. Algae die faster than they're eaten, and then voracious bacteria move in to consume them—a second bloom following the first, rapidly consuming oxygen. When the oxygen is gone, they shift their metabolic engines into an oxygen-free mode and keep on eating. The bacteria live without oxygen quite well, but they acidify and suffocate the sea for thousands of square miles around.

Productivity Bomb I: A sea of jellies

Overfishing and nutrient pollution react with each other, creating much worse problems than either might alone. So far, we have examined examples that carefully document each separately, but we have also seen them act together. We have seen the Productivity Bomb's results firsthand.

In the Soviet Union's dying days, the Black Sea faced the Bomb's full fury. In this small inland sea, porpoises had been hunted heavily for decades, and were only at about 10% of their historical numbers, so the small fish that were their normal prey flourished.[46] The abundant fish ate the zooplankton, small crustaceans that in turn normally graze on single-celled algae (phytoplankton) floating in the water. Too few porpoises led to many small fish, and too few zooplankton, and too many phytoplankton: ecologists consider this a classic tropic cascade. The first fuse of the Productivity Bomb—disrupted food chains—was lit.

Then, industrial farming along the rivers feeding into the Black Sea dumped a huge load of fertilizer into it, triggering massive blooms of floating algae. Massive reds tides plagued the Black Sea, creating dead zones that killed bottom-dwelling invertebrates and many fish.[47] The whole environment skewed toward algae, copepods, and anchovies until the little silver fish found themselves targeted by overactive fishing fleets. With anchovies gone, there was nothing left but algae and copepods.[48]

The copepod bonanza might have helped decrease the algal blooms. Enter a new species from elsewhere in the oceans—a comb jellyfish introduced into the Black Sea from the vast ballast tanks of intercontinental cargo ships.[49]

The jellyfish ate copepods aplenty, and in the depleted ecological setting of the Black Sea, there was nothing to hold back a jellyfish explosion. So they formed vast drifts in the cold water. It turns out the jellyfish swallowed juvenile anchovies too, preventing any fisheries recovery. This entire system, massively perturbed, had no way to correct itself. The result was leagues of barren water where once many fisheries had thrived. Barren, but for 3 tons of jellyfish in every square mile of ocean surface.[50]

The collapse of the Soviet Union cut off the flow of discarded fertilizer down the rivers leading to the Black Sea, and overfishing of anchovies ceased. The invasive jellies began to be devoured by a second invasive jellyfish that ate the first jellyfish. The Black Sea ecosystem gradually began to regain its balance, the grinding chain of destruction painfully rewinding. These changes were not simple: the repair bill included major shifts in agriculture, fishing, and porpoise protection. The first successes of all these efforts have begun to restore a functioning ecosystem in the Black Sea and show that the Productivity Bomb can be reversed.

Productivity Bomb II: Smothered corals

Tropical corals have long been half animal and half plant. This internal diversity lets them thrive when many creatures would starve. The coral animal eats as its ancestors did: sweeping its tentacles through the water to capture tiny creatures. Its protein diet is full of the phosphorous and nitrogen that plants need to grow, and so the animal passes these nutrients to algae that live placidly inside its own cells. The algae take the phosphorous and nitrogen and combine them with abundant sunlight to manufacture carbohydrate food. The interior algae transport these sugars to the coral host as rent payment.

It is a clever arrangement; the algae couldn't thrive in most tropical waters, because these oceans tend to be nutrient poor. So, the corals capture prey and use them for algae fertilizer. The corals cannot catch enough food to grow sturdy skeletons by themselves, and in turn they rely on the algae to drive their growth. Together they form a partnership that for hundreds of millions of years has survived under conditions that would starve each separately. What's more, the reefs they built shelter thousands of other species.

But this arrangement is deceptively fragile. Dump sewage and farm runoff into the water, consume most of the plant-eating fish, and watch a Productivity Bomb detonate. Fleshy green seaweeds—abundantly available once there is enough discarded fertilizer in the water—begin to grow quickly where they couldn't before. And where the herbivores have been removed by fishing, the algae can grow unchecked. Terry Hughes first saw this in Jamaica in the 1980s: millennia-old staghorn reefs and cities of mounding corals were carpeted and killed by weedy algae. Their survival hung by a thread.

That thread was a sea urchin named *Diadema,* a baseball-sized animal with venomous black spines the length of chopsticks. Its needle-sharp spines made it a bad food for fishermen, and they left it alone. But those fishermen did not leave the fish alone, and for decades the reefs of Jamaica had been stripped of algae-eating fish. Yet the coral reefs remained healthy and growing. The reason was that *Diadema* came out of its cracks and crevices at night and scoured the reef, eating all the algae they could find.

But then the thread broke. The *Diadema* were caught up in a massive disease outbreak that swept the Caribbean in a single year, killing millions or even billions of sea urchins. The last herbivores were gone, and no safeguards remained to defuse the Productivity Bomb. Algae swept over the reef, smothering corals until only a fraction of the original reef remained. Today, overfishing continues. The urchins have not fully recovered. And neither have the corals, relegated to a few thriving reefs, the rest scattered across their former domain in broken bone yards.

The slippery slope to slime

Jeremy Jackson, a saber-rattling marine ecologist (yes, such people exist) with long kinky orange hair pulled back in a ponytail, has long examined ocean problems through an unconventional lens. He has traveled the world, pointing out the oceans' fundamental collapse—the steady march through ever more devastating Productivity Bombs. Trained as a paleontologist, Jeremy helped create the field of historical ecology: comparing modern ecosystems to those of the past.[51] His "Reefs since Columbus" research was unveiled in 1996 and ever since, he's pointed to the demise of coral reefs as a genuine historical shift.[52] He and his colleagues call our present course "The slippery slope to slime," a wonderfully alliterative phrase describing the ocean's epochal shift from large fish and coastal gardens to a stew of bacteria, jellyfish, and tarlike algae.[53] The oceans of the future will be extremely productive—maybe even more so than today, in terms of raw biomass—but they'll leave much of the human race without the food we are used to, and perhaps with oceans devoted to microbes. The Productivity Bomb's most common physical product is slime.

Small, simple, soft, deep, blind, and weedy

Imagine an ocean seething with microbes, its red-brown surface weedy with algae. Vast dead zones and tides of neurotoxic water hamstring high-order productivity, wiping out the top half of the food chain while its base fills the world with toxic sludge. Waves crash on the beach, kicking up drifts of sticky green foam that cling to the sand. The clean salt air turns sickly toxic with the stench of decay. The ocean's delicate species have perished during the extremely hot days of summer or have expired in a bankruptcy of expensive skeletons.

Even the highly successful species in the ocean will find their comfortable niches altered by climate change, and like a pensioner on a fixed income, will not forever be able to absorb the metabolic tax of a warmer, more acidic ocean. The massive bolus of discarded fertilizer we put into the ocean, and the surgical precision with which we can cut out the very species that can quell the resulting ecological revolution, selects for a certain set of organisms that has always been in the oceans but hasn't been our favorite part.

We leave an ocean legacy of small, simple, soft, deep, blind, and weedy species. Curtains of jellyfish, lawns of microbes, a mud pile of worms, and perhaps a few lonely anglerfish in the deep sea. The oceans of the future will be alive—nothing will ever truly kill them—but what remains won't be the life we currently know.

Future extremes

The oceans are in a state of emergency, suffering from more ailments than science has remedies to offer.[54] If the policy "medicine" of reduced CO_2, better land management, fishing restraint, and protected areas isn't swiftly administered, the oceans will experience truly dire crises at ever-increasing rates.

All species will feel the pinch of changing temperature and acidity, from the depths of the sea where atmospheric CO_2 will eventually percolate, to the melting polar ice caps. Maybe hydrothermal vent shrimp won't notice much, bathed as they are in boiling poison. Maybe the broad abyssal plains will still flicker with bioluminescence. But great Antarctic whales will suffer when krill have to live without sea ice. The coldest species will suffer when the temperature rises a few degrees. The hottest corals, already living near their heat ceilings, will break through these ceilings. The shallowest intertidal species will bake at low tide and drown under storm surges. The oldest fish in the ocean may watch their descendants live progressively shorter, weaker lives. Even the deepest-ocean organisms—probably the least environmentally flexible species in the ocean—may find their homes awash in high acidity or altered currents.

Of course, in a few million years, conditions will improve. Nature is good at balancing itself out, after all. Previous massive alternations in the oceans have smoothed out, over geologic ages. And in that amount of time, the same time it has taken the human species to diverge from our ancestors, invent tools, and lift our eyes to wonder at the world around us, the planet and its diversity might recover.

Which is to say, over the long term the oceans don't need saving. *People* need saving. People will need to live through the next hundreds or thousands of years when the oceans are no longer the pantry of the world, no longer safe to swim in or sail across, toxic and wracked by ever-stronger storms. Hundreds of millions of people now live directly off the sea, and billions more benefit indirectly from marine life. Human society cannot simply wait out a centuries-long global Productivity Bomb or wait millennia for a return to "normal" ocean productivity. The fate of the oceans has become our fate too, and we out of easy ways to ensure that the future of the oceans is secure.

EPILOGUE A GRAND BARGAIN

A wooden skiff with a splintering hull and faded blue paint churns along under stormy skies. The coughing outboard engine takes it through warm, gentle Philippine waters. First green, then gray with reflected clouds, and finally dark with submerged reefs. Laden with heavy SCUBA gear, you swivel your head to the mossy crag overlooking the lagoon. Apo Island's steep volcanic slopes shelter a small village from strong South Pacific winds. A flotilla of fishing boats surrounds your skiff as the reedy pilot approaches your destination. All these fishing boats are a poor omen to start a dive; a sign of skittish and depleted marine life below.

Securing the mask over your face, you sink your teeth into the regulator's rubber and drop over the side. Sinking slowly through the clear, blue-tinged water, you get your first look at the reef. Your worries up above were blessedly premature. Countless fish swarm and twist through the reef, stunning in their colors and diversity. Bumphead wrasses sit in repose under coral heads, emerging on your approach. Three or four feet in length, they stake out territory and angrily posture. Friendlier fish flash past: green-and-purple parrotfish, platter-sized rays, and an armada of silver jacks that's grown legendary in local waters. This is a thriving reef, nowhere near decline.

Apo Island's reef is a rare and precious gem set in the heart of the Philippine archipelago—a region that has suffered heavy marine depletion. The large, healthy fish and the thriving corals sprawled beneath your feet sprouted from a single prescient decision. Decades ago, the island's small sheltered village decided to stop fishing on one part of the reef. It wasn't to

175

be touched for any reason, nor anything taken: it became what ecologists would term a marine protected area. The safe zone wasn't very big, and fishermen were allowed to fish right up to its invisible boundaries. Nevertheless, the impact of this small change on the reef was stupendous. Inside the protected zone, fish across the food chain could live a long time, growing to enormous size like the furious bumpheads. Instead of being taken immediately, they lived for years into maturity, cranking out millions of offspring. Many reef fish confine themselves to a home range the size of a swimming pool, so the lucky ones in protected areas can grow old and die without ever seeing a hook.[1] Outside the zone, existence is nasty, brutish, and short. Fish are snapped out of the water by the dense fishing fleet, and the fish stocks are all but gone. But the folks from Apo Island go home most nights with solid catches, taken from reefs just outside the protected zone. The protected area's immense productivity, left untouched, can replace what the fishermen take elsewhere.[2]

At this point in natural history, the oceans are cracking under our species' collective strain, but they have not truly broken. As we look forward to the year 2100, when the children of today will be grandparents, two broadly different futures swim into focus. One is the course we are on now, with CO_2 endlessly piling up in the atmosphere and oceans.[3] If by 2100 we're still pumping out carbon at our present rate or above (the top line in our accompanying graph), the oceans will not be salvageable, not returnable to their present state. By that point they will be too acidified, too warm, too high, and too stormy. And the amount of time that it will take for these climate changes to abate will be so long that by 2100, the damage is likely to be very long term.

But we can put ourselves onto a different CO_2 curve. If we do make the needed policy changes, then the oceans of the year 2100 may still be suffering—but they will not be ruined, and the long planet-cleansing process of reducing CO_2 will be underway. If emissions are tamed by 2050, then CO_2 in the atmosphere might begin to drop by 2100. The heat, storms, and acidification would then slowly begin to abate. It won't be pretty, but the point is that it will be getting better, not horribly worse, and the damage will be much shorter lived.

Scientists and the conservation community cannot put our society on a different CO_2 path. We are left with a grand challenge and a difficult bargain. The bargain is this: the economic forces, industries, and citizens of the world

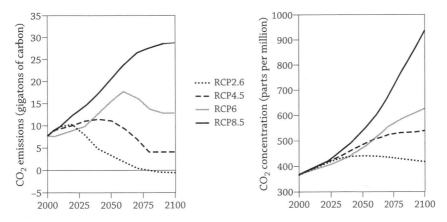

Predictions of CO_2 emissions (left) and atmospheric concentrations of CO_2 (right) based on various future scenarios of global response to climate change. The RCP 8.5 scenario (left figure, black solid line) is currently the most likely, as it represents no future controls on emissions. This scenario would lead to an exponential increase in oceanic CO_2 (right figure, black solid line) with serious impacts on ocean life from the year 2100 onward. Only if CO_2 emissions were to begin to decline by 2020 (for example, under the RCP 2.6 scenario, left figure dotted line) would CO_2 in the oceans begin to decrease by the year 2100. Intermediate scenarios (RCP 4.5 and RCP 6.0) would still see CO_2 in the oceans increasing for the foreseeable future

must work to do whatever it takes to stop the rise of CO_2 emissions by 2050 and return them to lower, tolerable levels by 2100. A different global energy source besides fossil fuels seems central to this success, but the shift does not have to be immediate. We have a generation to make this happen.

In return, the scientists and environmental engineers must do their best to save as much of the world's wild habitats, in the oceans and on land, into the next century, when conditions will improve. As many of the world's extreme and varied species as possible must be saved for the next century. When the climate begins to improve through the efforts and sacrifices and triumph of the global community, conservationists must promise to have a wild world ready to regrow.

Marine scientists know how to do their part. Protection and even recovery is already happening in places like Apo Island, where the big fish have returned, and California's Monterey Bay, where a single marine protected area became a foothold for sea otters to reconquer their old coastal haunts. Ocean life—extreme or not—is ready and able to thrive *for us*. The very same biological energy that makes the Productivity Bomb so terrifying can also repair the damage we've caused. The ocean itself is our single greatest tool when properly harnessed and leveraged. That tool sits ready, and we have a good idea how to use it through protecting habitats, employing sustainable fishing, guarding against discarded fertilizer and other coastal pollution, and kindling respect for the value of healthy oceans. No matter what we do, the seas of 2100 will teem with life. In lockstep with efforts to skid climate change to a halt, we still can choose to have the life of the sea be the whales, the tuna, the coral reefs, the jetting squid, the sea turtles, and smiling vaquitas.

NOTES

Prologue: The Epic Ocean

1. W. Whitman, *Leaves of Grass*, "Song of Myself," stanza 51.
2. http://www.lifesci.ucsb.edu/~biolum/organism/dragon.html.
3. Hoare, P. 2010. *The Whale: In Search of Giants in the Deep*. New York: Ecco.
4. Clarke, M. R. 1969. "A review of the systematics and ecology of oceanic squid." *Advances in Marine Biology* 4:91–300.
5. Roper, C. F., and K. J. Boss. 1982. "The giant squid." *Scientific American*, April.
6. Ellis, R. 1998. *The Search for the Giant Squid*. New York: Penguin.
7. Aoki, K., M. Amano, K. Mori, A. Kourogi, T. Kubodera, and N. Miyazaki. 2012. "Active hunting by deep diving sperm whales: 3D dive profiles and maneuvers during bursts of speed." *Marine Ecology Progress Series* 444:289–301.
8. Squid blood is blue, because it has no hemoglobin but instead uses a different protein called hemocyanin to transport oxygen. Hemocyanin is colorless when bound to oxygen, so it is hard to observe.
9. Whitehead, H. 2003. *Sperm Whales: Social Evolution in the Ocean*. Chicago: University of Chicago Press.
10. Ellis, *The Search for the Giant Squid*.

Chapter One: The Earliest

1. Kasting, J. F. 1993. "Earth's early atmosphere." *Science* 259:920–926.
2. Moseman, A. 2010. "Frost-covered asteroid suggests extraterrestrial origin for Earth's oceans." *Discover Magazine*, April 29.
3. Chyba, C., and C. Sagan. 1992. "Endogenous production, exogenous delivery and impact-shock synthesis of organic molecules: An inventory for the origins of life." *Nature* 355:125–132.
4. Knoll, A. H. 2004. *Life on a Young Planet: The First Three Billion Years of Evolution on Earth*. Princeton, NJ: Princeton University Press, p. 73.
5. Knoll, *Life on a Young Planet*, chapters 2, 3.
6. Tenenbaum, D. 2002. "When did life on Earth begin? Ask a rock." *Astrobiology Magazine*, October 14. http://astrobio.net/exclusive/293/when-did-life-on-earth

-begin-ask-a-rock; Olson, J. M. 2006. "Photosynthesis in the Archean era." *Photosynthesis Research* 88:109–117.

7. Rothschild, L. J. 2009. "Earth science: Life battered but unbowed." *Nature* 459:335–336;   http://www.nature.com/nature/journal/v459/n7245/full/459335a .html.

8. Wade, N. 2011. "Team claims it has found oldest fossils." *New York Times,* August 21; Wacey, D., M. R. Kilburn, M. Saunders, J. Cliff, and M. D. Brazier. "Microfossils of sulphur-metabolizing cells in 3.4-billion-year-old rocks of Western Australia." *Nature Geoscience* 4:698–702.

9. Olson, "Photosynthesis in the Archean era."

10. http://en.wikipedia.org/wiki/Great_Oxygenation_Event; Knoll, *Life on a Young Planet,* chapters 4, 5.

11. Olson, "Photosynthesis in the Archean era."

12. Buick, R. 2008. "When did oxygenic photosynthesis evolve?" *Philosophical Transactions of the Royal Society B* 363:2731–2743. doi: 10.1098/rstb.2008.0041.

13. See the exchange of letters at http://www.sciencemag.org/content/330/6005/ 754.2.full.pdf.

14. http://en.wikipedia.org/wiki/File:Oxygenation-atm-2.svg.

15. Lang, B. F., M. W. Gray, and G. Burger. 1999. "Mitochondrial genome evolution and the origin of eukaryotes." *Annual Review of Genetics* 33:351–397.

16. Knoll, *Life on a Young Planet,* chapters 6–8.

17. Gribaldo, S., A. M. Poole, V. Daubin, P. Forterre, and C. Brochier-Armanet. 2010. "The origin of eukaryotes and their relationship with the Archaea: Are we at a phylogenomic impasse?" *Nature Reviews Microbiology* 8:743–752.

18. Garrett, R., and H. P. Klenk. 2007. *Archaea: Evolution, Physiology, and Molecular Biology.* Oxford: Wiley and Sons.

19. Woese, C. R., O. Kandler, and M. L. Wheelis. 1990. "Towards a natural system of organisms: Proposal for the domains Archaea, Bacteria, and Eucarya." *Proceedings of the National Academy of Sciences, USA* 87: 4576–4579.

20. Barns, S. M., R. E. Fundyga, M. W. Jeffries, and N. R. Pace. 1994. "Remarkable Archaeal diversity detected in a Yellowstone National Park hot spring environment." *Proceedings of the National Academy of Sciences, USA* 91:1609–1613.

21. Blöchl, E., R. Rachel, S. Burggraf, D. Hafenbradl, H. W. Jannasch, and K. O. Stetter. 1997. "*Pyrolobus fumarii,* gen. and sp. nov., represents a novel group of Archaea, extending the upper temperature limit for life to 113 C." *Extremophiles* 1:14–21.

22. Narbonne, G. M. 2005. "The Ediacara biota: Neoproterozoic origin of animals and their ecosystems." *Annual Review of Earth and Planetary Sciences* 33:421–442.

23. Butterfield, N. J. 2011. "Terminal developments in Ediacaran embryology." *Science* 334:1655–1656.

24. Narbonne, "The Ediacara biota." See also http://en.wikipedia.org/wiki/ Ediacaran_biota.

25. Gould, S. J. 1990. *Wonderful Life: The Burgess Shale and the Nature of History.* New York: W. W. Norton and Company; Knoll, *Life on a Young Planet,* chapter 11.

26. Gould, *Wonderful Life,* chapter 3.

27. Knoll, *Life on a Young Planet,* pp. 192–193; see also Gould, *Wonderful Life,* pp. 124–136.

28. Ho, S. 2008. "The Molecular Clock and estimating species divergence." *Nature Education* 1.

29. Wray, G., J. S. Levinton, and L. Shapiro. 1996. "Molecular evidence for deep Precambrian divergences among metazoan phyla." *Science* 274:568–573.

30. Jensen, S. 2003. "The Proterozoic and earliest Cambrian trace fossil record; Patterns, problems and perspectives." *Integrative and Comparative Biology* 43:219–228; http://en.wikipedia.org/wiki/Trace_fossil.

31. Benton, M. J. 2008. *The History of Life: A Very Short Introduction.* Oxford: Oxford University Press, chapter 3.

32. Morris, S. C., and J. B. Caron. 2012. "*Pikaia gracilens* Walcott, a stem-group chordate from the Middle Cambrian of British Columbia." *Biological Reviews* 87:480–512.

33. Gould, *Wonderful Life,* pp. 312–323.

34. Whittington, H. B. 1975. "The enigmatic animal *Opabinia regalis,* Middle Cambrian Burgess Shale, British Columbia." *Philosophical Transactions of the Royal Society of London B* 271:1–43.

35. Gould, *Wonderful Life,* pp.124–136; http://en.wikipedia.org/wiki/Opabinia.

36. Gould, *Wonderful Life,* pp. 124–125.

37. Gould, *Wonderful Life,* p. 51.

38. Gould, *Wonderful Life,* p. 25.

39. Gould, *Wonderful Life,* pp. 124–136.

40. Gould, *Wonderful Life,* p. 47.

Chapter Two: The Most Archaic

1. http://en.wikipedia.org/wiki/Volkswagen_air_cooled_engine.

2. Braddy, S. J., M. Poschmann, and O. E. Tetlie. 2008. "Giant claw reveals the largest ever arthropod." *Biology Letters* 4:106–109.

3. Lieberman, B. S. 2002. "Phylogenetic analysis of some basal early Cambrian trilobites, the biogeographic origins of the eutrilobita, and the timing of the Cambrian radiation." *Journal of Paleontology* 76:692–708. doi: 10.1666/0022-3360.

4. Fortey, Richard. 2000. *Trilobite! Eyewitness to Evolution.* New York: Knopf Doubleday, p. 214.

5. See Fortey, *Trilobite!* chapter 9.

6. Clarkson, E., R. Levi-Setti, and G. Horváth. 2006. "The eyes of trilobites: The oldest preserved visual system." *Arthropod Structure and Development* 35:247–259.

7. Fortey, R., and B. Chatterton. 2003. "A Devonian trilobite with an eyeshade." *Science* 301:1689. doi: 10.1126/science.1088713.

8. Towe, K. M. 1973. "Trilobite eyes: Calcified lenses in vivo." *Science* 179:1007–1009. doi: 10.1126/science.179.4077.1007.

9. Fortey, *Trilobite!* p. 241.

10. Owens, R. M. 2003. "The stratigraphical distribution and extinctions of Permian trilobites." *Special Papers in Palaeontology* 70:377–397.

11. See a fine 6-foot specimen from British Columbia featured prominently at http://www.bcfossils.ca/.

12. Cook, T. A. 1979. *The Curves of Life: Being an Account of Spiral Formations and Their Application to Growth in Nature, to Science, and to Art: With Special Reference to the Manuscripts of Leonardo da Vinci.* Mineola, NY: Courier Dover Publications.

13. Collins, D. H., and P. Minton. 1967. "Siphuncular tube of *Nautilus*." *Nature* 216:916–917.

14. Collins and Minton, "Siphuncular tube."

15. Boardman, R. S., A. H. Cheetham, and A. J. Rowell. 1987. *Fossil Invertebrates*. Oxford: Blackwell, p. 345.

16. Castro, P., and M. Huber. 2000. *Marine Biology,* third edition. Boston: McGraw-Hill, p. 350.

17. *Hamlet,* Act 2, Scene 2.

18. Fortey, R. 2012. *Horseshoe Crabs and Velvet Worms: The Story of the Animals and Plants That Time Has Left Behind.* New York: Knopf.

19. Avise, J. C., W. S. Nelson, and H. Sugita. 1994. "A speciational history of 'living fossils': Molecular evolutionary patterns in horseshoe crabs." *Evolution* 48:1986–2001.

20. http://www.ceoe.udel.edu/horseshoecrab/research/eye.html.

21. It diverged from the other horseshoe crabs 27–70 million years ago, but it does not have a fossil record distinct enough to track it's fossil history. Avise et al., "A speciational history of 'living fossils.'"

22. Størmer, L. 1952. "Phylogeny and taxonomy of fossil horseshoe crabs." *Journal of Paleontology* 26:630–640.

23. http://www.sciencedaily.com/releases/2008/02/080207135801.htm.

24. Balon, E. K., M. N. Bruton, and H. Fricke. 1988. "A fiftieth anniversary reflection on the living coelacanth, *Latimeria chalumnae:* Some new interpretations of its natural history and conservation status." *Environmental Biology of Fishes* 23:241–280.

25. http://vertebrates.si.edu/fishes/coelacanth/coelacanth_wider.html.

26. Quoted in Balon et al., "A fiftieth anniversary," p. 243.

27. Balon et al., "A fiftieth anniversary."

28. Eilperin, J. 2012. *Demon Fish: Travels through the Hidden World of Sharks.* New York: Anchor Books, p. 25.

29. Eilperin, *Demon Fish.*

30. Kalmijn, A. J. 2000. "Detection and processing of electromagnetic and near-field acoustic signals in elasmobranch fishes." *Philosophical Transactions of the Royal Society of London B* 355:1135–1141.

31. For an engaging scientific narrative about pioneering but simple experiments in shark electrosensing, see Kalmijn, A. J. 1971. "The electric sense of sharks and rays." *Journal of Experimental Biology* 55:371–383.

32. http://www.ams.org/samplings/feature-column/fcarc-pagerank.

33. Paleontological Society. "Fossil shark teeth." Paleosoc.org. http://www.paleosoc .org/Fossil_Shark_Teeth.pdf.

34. Vampire bat teeth rate highly, too, scalpels able to slice a vein without a victim noticing. Feldhamer, G., L. C. Drickhamer, S. H. Vessey, J. F. Merritt, and C. Krajewski. 2007. *Mammalogy: Adaptation, Diversity, Ecology.* Baltimore: Johns Hopkins University Press, p. 63.

35. Kemp, N. E., and J. H. Park. 1974. "Ultrastructure of the enamel layer in developing teeth of the shark *Carcharhinus menisorrah.*" *Archives of Oral Biology* 19:633–644.

36. Kemp, N. E. 1985. "Ameloblastic secretion and calcification of the enamel layer in shark teeth." *Journal of Morphology* 184:215–230.

37. Grogan, E. D., R. Lund, and E. Greenfest-Allen. 2004. "The origin and relationships of early chondrichthyans." In J. C. Carrier, J. A. Musick, and M. R. Heithaus (eds.). *Biology of Sharks and Their Relatives*. Boca Raton, FL: CRC Press, pp. 3–32. For an online alternative, see http://www.elasmo-research.org/education/evolution/earliest.htm.

38. Miller, R. F., R. Cloutier, and S. Turner. 2003. "The oldest articulated chondrichthyan from the Early Devonian period." *Nature* 425:501–504.

39. http://en.wikipedia.org/wiki/Cladoselache.

40. Vanessa Jordan, Florida Museum of Natural History. http://www.flmnh.ufl.edu/fish/gallery/descript/goblinshark/goblinshark.html.

41. Compagno, L.J.V. 1977. "Phyletic relationships of living sharks and rays." *American Zoologist* 17:303–322.

42. http://science.nationalgeographic.com/science/prehistoric-world/permian-extinction/.

43. Shen S.-Z., J. L. Crowley, Y. Wang, S. A. Bowring, D. H. Erwin, et al. 2011. "Calibrating the end-Permian mass extinction." *Science* 334:1367–1372. doi: 10.1126/science.1213454.

44. Benton, M. J., and R. J. Twitchett. 2003. "How to kill (almost) all life: The end-Permian extinction event." *Trends in Ecology and Evolution* 18:358–365.

45. Payne, J. L., D. J. Lehrmann, J. Wei, M. J. Orchard, D. P. Schrag, and A. H. Knoll. 2004. "Large perturbations of the carbon cycle during recovery from the end-Permian extinction." *Science* 305:506–509.

46. Dean, M. N., C. D. Wilga, and A. P. Summers. 2005. "Eating without hands or tongue: Specialization, elaboration and the evolution of prey processing mechanisms in cartilaginous fishes." *Biology Letters* 1:357–361. doi: 10.1098/rsbl.2005.0319 1744 957X.

47. http://www.flmnh.ufl.edu/fish/sharks/fossils/megalodon.html.

48. Gottfried, M. D., L.J.V. Compagno, and S. C. Bowman. 1996. "Size and skeletal anatomy of the giant megatooth shark *Carcharodon megalodon*." In A. P. Klimley and D. G. Ainley (eds.). *Great White Sharks: The Biology of* Carcharodon carcharias. San Diego: Academic Press, pp. 55–89.

49. http://www.flmnh.ufl.edu/fish/sharks/fossils/megalodon.html.

50. Botella, H., P.C.J. Donoghue, and C. Martínez-Pérez. 2009. "Enameloid microstructure in the oldest known chondrichthyan teeth." *Acta Zoologica* 90(supplement):103–108.

51. Barnosky, A. D., N. Matzke, S. Tomiya, G. O. Wogan, B. Swartz, et al. 2011. "Has the Earth's sixth mass extinction already arrived?" *Nature* 471:51–57.

52. Holder, M. T., M. V. Erdmann, T. P. Wilcox, R. L. Caldwell, and D. M. Hillis. 1999. "Two living species of coelacanths?" *Proceedings of the National Academy of Sciences, USA* 96:12616–12620.

53. Saunders, W. B. 2010. "The species of nautilus." In W. B. Sanders and N. H. Landers (eds.). *Nautilus: The Biology and Paleobiology of a Living Fossil*. Dordrecht: Springer, pp. 35–52.

54. http://www.washingtonpost.com/wp-dyn/content/article/2005/06/09/AR2005060901894.html.

Chapter Three: The Smallest

1. http://www.sciencedaily.com/releases/2008/06/080603085914.htm; http://en.wikipedia.org/wiki/Human_microbiome.

2. http://en.wikipedia.org/wiki/Microscope.

3. Pasteur, L. 1878. "The germ theory and its applications to medicine and surgery." Read before the French Academy of Sciences, April 29, 1878. *Comptes Rendus de l' Academie des Sciences* 86: 1037–1043.

4. Darwin, C. 1845. *Voyage of the Beagle,* second edition. London: John Murray, p. 519.

5. Pomeroy, L. R. 1974. "The ocean's food web, a changing paradigm." *BioScience* 24:499–504; see also the original counting methodology reviewed in Hobbie, J. E., R. J. Daley, and S. Jasper. 1977. "Use of Nuclepore filters for counting bacteria by fluorescence microscopy." *Applied and Environmental Microbiology* 33:1225–1228.

6. Pomeroy, L. R., P.J.I. Williams, F. Azam, and J. E. Hobbie. 2007. "The microbial loop." *Oceanography* 20(2):28–33.

7. You probably need proof of this, too. We certainly did. The ocean's 10^{29} bacteria, each one millionth of a meter long, would stretch out to 10^{20} kilometers or about 10,000,000 light-years. The Milky Way's circumference measures roughly 300,000 light-years. Adapted from George Somero and Mark Denny, Oceanic Biology lecture, Hopkins Marine Station, Pacific Grove, CA, February 2012.

8. Ducklow, H. W. 1983. "Production and fate of bacteria in the oceans." *BioScience* 33:494–501.

9. Pomeroy et al., "The microbial loop," pp. 28–33.

10. Cho, B. C., and F. Azam. 1988. "Major role of bacteria in biogeochemical fluxes in the ocean's interior." *Nature* 332:441–443.

11. Johnson, P. W., and J. M. Seiburth. 1979. "Chroococcoid cyanobacteria in the sea: A ubiquitous and diverse phototrophic biomass." *Limnology and Oceanography* 24:928–935.

12. See the lively report of a talk by Penny Chisolm in Beardsley, T. M. 2006. "Metagenomics reveals microbial diversity." *BioScience* 56:192–196.

13. Also see this nice report from National Public Radio's Science Friday: http://www.npr.org/templates/story/story.php?storyId=91448837.

14. Penny Chisholm, personal communication, March 2012.

15. Campbell, N. A., J. B. Reece, M. R. Taylor, E. J. Simon, and J. L. Dickey. 2006. *Biology: Concepts and Connections.* New York: Benjamin Cummings.

16. http://microbewiki.kenyon.edu/index.php/Prochlorococcus_marinus.

17. http://en.wikipedia.org/wiki/Mycoplasma_genitalium.

18. http://www.scientificamerican.com/article.cfm?id=gulf-oil-eating-microbes-slide-show.

19. See the recent review in Hartmann, M., C. Grob, G. A. Tarran, A. P. Martin, P. H. Burkill, et al. 2012. "Mixotrophic basis of Atlantic oligotrophic ecosystems." *Proceedings of the National Academy of Sciences, USA* 109:5756–5760.

20. http://en.wikipedia.org/wiki/Dissolved_organic_carbon; http://en.wikipedia.org/wiki/Microbial_loop; Pomeroy et al., "The microbial loop," pp. 30–31.

21. Stone, R. 2010. "Marine biogeochemistry: The invisible hand behind a vast carbon reservoir." *Science* 328:1476–1477.

22. Zubkov, M. V., M. A. Sleigh, and P. H. Burkill. 2001. "Heterotrophic bacterial turnover along the 20°W meridian between 59°N and 37°N in July 1996." *Deep-Sea Research Part II: Topical Studies in Oceanography* 48:987–1001.

23. If a billion bacteria weigh 100 nanograms and there are 10^{29} bacteria, a little math gives you this figure.

24. Azam, F., T. Fenchel, J. G. Field, J. S. Gray, L. A. Meyer-Reil, and F. Thingstad. 1983. "The ecological role of water-column microbes in the sea." *Marine Ecology Progress Series* 10:257–263.

25. Pomeroy et al., "The microbial loop," p. 28.

26. We attempt a small academic balancing act here. Many of the most entertaining microbes are part of a group called the Archaea (see Chapter 1). They are distinct from what scientists now call the Eubacteria. As Azam and Malfatti put it, we should call some of them Archaea and others bacteria, but for the sake of word economy, and some sanity, we are calling all of them bacteria or sometimes microbes. See Azam, F., and F. Malfatti. 2007. "Microbial structuring of marine ecosystems." *Nature Reviews Microbiology* 5:782–791.

27. Bratbak, G., and M. Heldal. 2000. "Viruses rule the waves—The smallest and most abundant members of marine ecosystems." *Microbiology Today* 27:171–173.

28. http://www.sciencemag.org/content/335/6072/1035.full.

29. Bratbak and Heldal, "Viruses rule the waves."

30. Suttle, C. A. 2007. "Marine viruses—Major players in the global ecosystem." *Nature Reviews Microbiology* 5:801–812.

31. http://researcharchive.calacademy.org/research/scipubs/pdfs/v56/proccas_v56_n06_Suppl.pdf.

32. Weiss, K. 2006. "A primeval tide of toxins." *Los Angeles Times*, July 30. http://articles.latimes.com/2006/jul/30/local/la me ocean30jul30.

33. Palumbi, S. R. 2001. *The Evolution Explosion*. New York: W. W. Norton.

34. Rohwer, F., and R. V. Thurber. 2009. "Viruses manipulate the marine environment." *Nature* 459:207–212.

35. Bidle, K. D., L. Haramaty, J.B.E. Ramos, and P. Falkowski. 2007. Viral activation and recruitment of metacaspases in the unicellular coccolithophore, *Emiliania huxleyi*." *Proceedings of the National Academy of Sciences, USA* 104:6049–6054.

36. Rohwer and Thurber, "Viruses manipulate the marine environment."

37. Frada, M., I. Probert, M. Allen, W. Wilson, and C. de Vargas, C. 2008. "The 'Cheshire Cat' escape strategy of the coccolithophore *Emiliania huxleyi* in response to viral infection." *Proceedings of the National Academy of Sciences, USA* 105:15944.

38. Palumbi, *Evolution Explosion*.

39. Meyer, K. M., M. Yu, A. B. Jost, B. M. Kelley, and J. L. Payne. 2010. "δ^{13}C evidence that high primary productivity delayed recovery from end-Permian mass extinction." *Earth and Planetary Science Letters* 302:378–384. doi: 10.1016/j.epsl.2010.12.033.

40. Fenchel, T. 2008. "The microbial loop—25 years later." *Journal of Experimental Marine Biology and Ecology* 366:99–103.

Chapter Four: The Deepest

1. Beebe, W. 1935. *Half Mile Down*. London: John Lane, p. 102. Quote is from p. 112.

2. Beebe, *Half Mile Down,* p. 147.

3. It is not clear whether Beebe ever said this or merely thought it. Beebe, *Half Mile Down,* p. 100.

4. Beebe was director of the Department of Tropical Research at the New York Zoological Society.

5. There is occasional vegetable matter—such as logs and tree trunks. For example, see Turner, R. D. 1973. "Wood-boring bivalves, opportunistic species in the deep sea." *Science* 180:1377–1379. doi: 10.1126/science.180.4093.1377.

6. Lonsdale, P. 1977. "Clustering of suspension-feeding macrobenthos near abyssal hydrothermal vents at oceanic spreading centers." *Deep Sea Research* 24:857–863.

7. Tivey, M. K. 1998. "How to build a black smoker chimney." *Oceanus*, December 1998. http://www.whoi.edu/oceanus/viewArticle.do?id=2400.

8. For a general overview, see http://www.csa.com/discoveryguides/vent/review2 .php.

9. Jannasch, H. W. 1985. "The chemosynthetic support of life and the microbial diversity at deep-sea hydrothermal vents." *Proceedings of the Royal Society of London B* 225:277–297; Felbeck, H., J. J. Childress, and G. N. Somero. 1981. "Calvin-Benson cycle and sulphide oxidation enzymes in animals from sulphide-rich habitats." *Nature* 293:291. doi: 10.1038/293291a0.

10. Mascarelli, A. 2009. "Dead whales make for an underwater feast." *Audubon Magazine,* November–December. http://www.audubonmagazine.org/articles/nature/ dead-whales-make-underwater-feast.

11. http://www.mnh.si.edu/onehundredyears/featured_objects/Riftia.html.

12. Cavanaugh, C. M., S. L. Gardiner, M. L. Jones, H. W. Jannasch, and J. B. Waterbury. 1981. "Prokaryotic cells in the hydrothermal vent tube worm *Riftia pachyptila* Jones: Possible chemoautotrophic symbionts." *Science* 213: 340–342.

13. Bailly, X., and S. Vinogradov. 2005. "The sulfide binding function of annelid hemoglobins: Relic of an old biosystem?" *Journal of Inorganic Biochemistry* 99:142–150.

14. Childress, J. J., and C. R. Fisher. 1992. "The biology of hydrothermal vent animals: Physiology, biochemistry and autotrophic symbioses." *Annual Review of Oceanography and Marine Biology* 30:337–441.

15. Jones, M. L., and S. L. Gardiner. 1989. "On the early development of the vestimentiferan tube worm *Ridgeia* sp. and observations on the nervous system and trophosome of *Ridgeia* sp. and *Riftia pachyptila.*" *Biological Bulletin* 177:254–276.

16. Nussbaumer, A. D., C. R. Fisher, and M. Bright. 2006. "Horizontal endosymbiont transmission in hydrothermal vent tubeworms." *Nature* 441:345–348.

17. Lutz, R. A., T. M. Shank, D. J. Fornari, R. M. Haymon, M. D. Lilley, et al. 1994. "Rapid growth at deep-sea vents." *Nature* 371:663–664.

18. http://www.sciencedaily.com/releases/2000/02/000203075002.htm.

19. The actual number is hard to determine and increases with every new expedition. Baker and colleagues describe the work for the Census of Marine Life in Baker, M. C., E. Z. Ramirez-Llodra, P. A. Tyler, C. R. German, A. Boetius, et al. 2010. "Biogeography, ecology, and vulnerability of chemosynthetic ecosystems in the deep sea." In A. D. McIntyre (ed.). *Life in the World's Oceans.* Oxford: Blackwell; http://blogs .nature.com/news/2012/01/hydrothermal-vents-host-a-bonanza-of-new-species .html.

20. http://www.livescience.com/17715-yeti-crabs-antarctic-vents.html.

21. Castro, P., and M. Huber. 2000. *Marine Biology,* third edition. Boston: McGraw-Hill, p. 338.

22. Butman, C. A., J. T. Carlton, and S. R. Palumbi. 1996. "Whales don't fall like snow: Reply to Jelmert." *Conservation Biology* 10:655–656.

23. http://www.emagazine.com/magazine-archive/whale-falls.

24. A fuller description of the drama of a whale fall is in Little, C.T.S. 2010. "Life at the bottom: The prolific afterlife of whales." *Scientific American,* February. doi: 10 .1038/scientificamerican0210-78. http://www.scentificamerican.com/article.cfm?id= the-prolific-afterlife-of-whales.

25. Little, "Life at the bottom."

26. Little, "Life at the bottom"; see also http://www.mbari.org/news/news_ releases/2010/whalefalls/whalefalls-release.html.

27. Butman, C. A., J. T. Carlton, and S. R. Palumbi. 1995. "Whaling effects on deep -sea biodiversity." *Conservation Biology* 9:462–464.

28. Smith, C., and A. Baco. 2003. "The ecology of whale falls at the deep sea floor." *Annual Review of Oceanography and Marine Biology* 41:311–354.

29. http://www.mbari.org/news/news_releases/2002/dec20_whalefall.html.

30. http://www.mbari.org/twenty/osedax.htm.

31. http://www.mbari.org/news/news_releases/2002/dec20_whalefall.html.

32. Rouse, G. W., S. K. Goffredi, and R. C. Vrijenhoek. 2004. "*Osedax:* Bone-eating marine worms with dwarf males." *Science* 305:668–671. doi: 10.1126/science.1098650; http://www.audubonmagazine.org/truenature/truenature0911.html.

33. Rouse et al., "*Osedax:* Bone-eating marine worms."

34. Rouse, G. W., K. Worsaae, S. B. Johnson, W. J. Jones, and R. C. Vrijenhoek. 2008. "Acquisition of dwarf male 'harems' by recently settled females of *Osedax roseus* n. sp. (Siboglinidae; Annelida)." *Biology Bulletin* 214:67–82. doi: 10.2307/25066661.

35. http://www.nature.com/news/2010/101206/full/news.2010.651.html.

36. Rouse, G. W., S. K. Goffredi, S. B. Johnson, and R. C. Vrijenhoek. 2011. "Not whale-fall specialists, *Osedax* worms also consume fishbones." Biology Letters 7:736–739. doi: 10.1098/rsbl.2011.0202.

37. http://en.wikipedia.org/wiki/Boyle's_law.

38. Kooyman, G. L. 2009. "Diving physiology." In W. F. Perrin, B. Wursig, and J.G.M. Thewissen (eds.). *Encyclopedia of Marine Mammals.* San Diego, CA: Academic Press, pp. 327–332.

39. For a nice romp through deep-sea effects, see http://discovermagazine.com/ 2001/aug/featphysics.

40. http://discovermagazine.com/2001/aug/featphysics.

41. http://en.wikipedia.org/wiki/Saturated_fat.

42. Cossins, A. R., and A. G. Macdonald. 1989. "The adaptation of biological membranes to temperature and pressure: Fish from the deep and cold." *Journal of Bioenergetics and Biomembranes* 21:115–135.

43. For example, see the relationships between deep sea rattails (sometimes more elegantly called grenadiers) and their shallow water relatives: Morita, T. 2004. "Studies on molecular mechanisms underlying high pressure adaptation of α-actin from deep-sea fish." *Bulletin of the Fisheries Research Agency* 13:35–77.

44. Oliver, T. A., D. A. Garfield, M. K. Manier, R. Haygood, G. A. Wray, and S. R. Palumbi. 2010. "Whole-genome positive selection and habitat-driven evolution in a shallow and a deep-sea urchin." *Genome Biology and Evolution* 2:800.

45. Kaariainen, J., and B. Bett. 2010. "Evidence for benthic body size miniaturization in the deep sea." *Journal of the Marine Biological Association of the UK* 86:1339–1345.

46. A fairly famous photograph shows this: http://korovieva.files.wordpress.com/2010/05/giantisopods_doritos1.jpg.

47. http://scienceblogs.com/deepseanews/2007/04/from_the_desk_of_zelnio_bathyn.php.

48. http://davehubbleecology.blogspot.com/2012/02/antarctic-sea-spiders-polar-or-abyssal.html

49. Fisher, C. R., I. A. Urcuyo, M. A. Simpkins, and E. Nix. 1997. "Life in the slow lane: Growth and longevity of cold-seep vestimentiferans." *Marine Ecology* 18:83–94.

50. Woods, H. A., A. L. Moran, C. P. Arango, L. Mullen, and C. Shields. 2008. "Oxygen hypothesis of polar gigantism not supported by performance of Antarctic pycnogonids in hypoxia." *Proceedings of the Royal Society of London B* 276:1069–1075.

51. Timofeev, S. F. 2001. "Bergmann's Principle and deep-water gigantism in marine crustaceans." *Biology Bulletin* 28:646–650. http://www.springerlink.com/content/w40861j17433662t/.

52. Castro and Huber, *Marine Biology,* pp. 345–347.

53. Castro and Huber, *Marine Biology,* pp. 345–347.

54. http://www.tonmo.com/science/public/giantsquidfacts.php.

55. http://en.wikipedia.org/wiki/Colossal_squid#Largest_known_specimen.

56. Sweeney, M. J., and C.F.E. Roper. 2001. *Records of* Architeuthis *Specimens from Published Reports.* Washington, DC: National Museum of Natural History, Smithsonian Institution. See http://invertebrates.si.edu/cephs/archirec.pdf.

57. http://www.tonmo.com/science/public/giantsquidfacts.php.

58. How could we actually know for sure?

59. Winkelmann, I., P. F. Campos, J. Strugnell, Y. Cherel, P. J. Smith, et al. 2013. "Mitochondrial genome diversity and population structure of the giant squid *Architeuthis:* Genetics sheds new light on one of the most enigmatic marine species. *Proceedings of the Royal Society of London B* 280:1759.

60. http://invertebrates.si.edu/giant_squid/page2.html.

61. Winkelmann et al., "Mitochondrial genome diversity and population structure of the giant squid."

62. Mesnick, S. L., B. L. Taylor, F. I. Archer, K. K. Martien, S. E. Treviño, et al. 2011. "Sperm whale population structure in the eastern and central North Pacific inferred by the use of single-nucleotide polymorphisms, microsatellites and mitochondrial DNA." *Molecular Ecology Resources* 11(supplement): 278–298.

63. http://squid.tepapa.govt.nz/exhibition.

64. http://news.nationalgeographic.com/news/2007/02/070222-squid-pictures.html.

65. http://news.nationalgeographic.com/news/2005/09/0927_050927_giant_squid.html.

66. Most of the following discussion is summarized in Haddock, S.H.D., M. A. Moline, and J. F. Case. 2010. "Bioluminescence in the sea." *Annual Reviews of Marine Science* 2:443–493.

67. Castro and Huber, *Marine Biology*, p. 341.

68. Castro and Huber, *Marine Biology*, p. 342.

69. Called dinoflagellates; see Haddock et al., "Bioluminescence in the sea," p. 465.

70. Haddock et al., "Bioluminescence in the sea."

71. See the remarkable book Piestch, T. 2009. *Oceanic Anglerfishes: Extraordinary Diversity in the Deep Sea.* Berkeley: University of California Press.

72. Piestch, *Oceanic Anglerfishes,* p. 7.

73. But the anglerfish was smiling. Piestch, *Oceanic Anglerfishes,* p. 262–263.

74. Widder, E. A., M. I. Latz, P. J. Herring, and J. F. Case. 1984. "Far-red bioluminescence from two deep-sea fishes." *Science* 225:512–514.

75. Herring, P. J., and C. Cope. 2005. "Red bioluminescence in fishes: On the suborbital photophores of *Malacosteus, Pachystomias* and *Aristostomias.*" *Marine Biology* 148:383–394.

76. Hunt, D. M., S. D. Kanwaljit, J. C. Partridge, P. Cottrill, and J. K. Bowmaker. 2001. "The molecular basis for spectral tuning of rod visual pigments in deep-sea fish." *Journal of Experimental Biology* 204:3333–3344.

77. Beebe, *Half Mile Down,* p. 225.

Chapter Five: The Shallowest

1. Gallien, W. B. 1986. "A comparison of hydrodynamic forces on two sympatric sea urchins: Implications of morphology and habitat." Thesis, University of Hawaii, Honolulu. See also Denny, M., and B. Gaylord. 1996. "Why the urchin lost its spines: Hydrodynamic forces and survivorship in three echinoids." *Journal of Experomental Biology* 199:717–729.

2. Stephenson, T. A., and A. Stephenson. 1972. *Life between Tidemarks on Rocky Shores.* San Francisco: W. H. Freeman and Company.

3. Stephenson, T. A., and A. Stephenson. 1949. "The universal features of zonation between tide-marks on rocky coasts." *Journal of Ecology* 37:289–305. The quote is from p. 303.

4. Castro, P., and M. Huber. 2000. *Marine Biology,* third edition. Boston: McGraw-Hill, p. 225.

5. Castro and Huber, *Marine Biology,* p 228.

6. Glynn, P. W. 1997. "Bioerosion and coral-reef growth: A dynamic balance." In C. Birkeland (ed.). *Life and Death of Coral Reefs.* New York: Springer, pp. 68–95.

7. Sokolova, I. M., and H. O. Pörtner. 2001. "Physiological adaptations to high intertidal life involve improved water conservation abilities and metabolic rate depression in *Littorina saxatilis.*" *Marine Ecology Progress Series* 224:171–186.

8. Garrity, S. D. 1984. "Some adaptations of gastropods to physical stress on a tropical rocky Shore." *Ecology* 65:559–574.

9. Curtis L. A. 1987. "Vertical distribution of an estuarine snail altered by a parasite." *Science* 235:1509–1511.

10. http://en.wikipedia.org/wiki/Salt_marsh.

11. Bertness, M. D. 1998. *Atlantic Shorelines: Natural History and Ecology.* Sunderland, MA: Sinauer Press.

12. Bertness, M. D. 1984. "Ribbed mussels and *Spartina alterniflora* production in a New England salt marsh." *Ecology* 65:1794–1807.

13. Castro and Huber, *Marine Biology,* p. 251.

14. http://www.flmnh.ufl.edu/fish/southflorida/mangrove/adaptations.html.

15. http://www.sms.si.edu/irlspec/Mangroves.htm.

16. Mann, K. H. 2000. "Estuarine benthic systems." In K. H. Mann (ed.). *Ecology of Coastal Waters with Implications for Management.* Oxford: Blackwell, pp. 118–135.

17. Reef, R., I. C. Feller, and C. E. Lovelock. 2010. "Nutrition of mangroves." *Tree Physiology* 30:1148–1160.

18. For example, see Lutz, P. 1997. "Salt, water and pH balance in the sea turtle." In P. Lutz and J. Musick (eds.). *The Biology of Sea Turtles.* Boca Raton, FL: CRC Press, pp. 343–361.

19. http://miami-dade.ifas.ufl.edu/documents/MangroveFactSheet.pdf.

20. Scholander, P. F. 1968. "How mangroves desalinate water." *Physiologia Plantarum* 21:251–261.

21. http://www.nhmi.org/mangroves/phy.htm.

22. Evans, L. S., and A. Bromberg. 2010. "Characterization of cork warts and aerenchyma in leaves of *Rhizophora mangle* and *Rhizophora racemosa.*" *Journal of the Torrey Botanical Society* 137:30–38.

23. Mumby, P. J., A. J. Edwards, J. E. Arias-González, K. C. Lindeman, P. G. Blackwell, et al. 2004. "Mangroves enhance the biomass of coral reef fish communities in the Caribbean." *Nature* 427:533–536.

24. Harris, V. A. 1960. "On the locomotion of the mudskipper *Periophthalmus koelreuteri* (Pallas): Gobiidae." *Proceedings of the Zoological Society of London* 134:107–135. doi: 10.1111/j.1469-7998.1960.tb05921.x.

25. Harris, "On the locomotion of the mudskipper."

26. Graham, J. B. (ed.). 1997. *Air-Breathing Fishes. Evolution, Diversity and Adaptation.* San Diego, CA: Academic Press.

27. Castro and Huber, *Marine Biology,* p. 254.

28. http://en.wikipedia.org/wiki/Mudskipper.

29. Morris, R. H., D. P. Abbott, and E. C. Haderlie. 1980. *Intertidal Invertebrates of California.* Palo, Alto, CA: Stanford University Press.

30. Ricketts, E. F., and J. Calvin. 1985. *Between Pacific Tides,* fifth edition. Palo Alto, CA: Stanford University Press.

31. Castro and Huber. *Marine Biology,* p. 228.

32. Morris et al., *Intertidal Invertebrates of California.*

33. http://www.washington.edu/research/pathbreakers/1969g.html.

Chapter Six: The Oldest

1. Simon, S. L., and W. L. Robison. 1997. "A compilation of nuclear weapons test detonation data for US Pacific Ocean tests." *Health Physics* 73:258–264.

2. Cailliet, G., and A. Andrews. 2008. "Age-validated longevity of fishes: Its importance for sustainable fisheries." In K. Tsukamoto, T. Kawamura, T. Takeuchi, T. D.

Beard Jr., and M. J. Kaiser (eds.). *5th World Fisheries Congress 2008: Fisheries for Global Welfare and Environment*. Tokyo: TERRAPUB, pp. 103–120.

3. http://www.afsc.noaa.gov/REFM/age/FAQs.htm.

4. Brodie, P. F. 1971. "A reconsideration of aspects of growth, reproduction, and behavior of the white whale (*Delphinapterus leucas*), with reference to the Cumberland Sound, Baffin Island, population." *Journal of the Fisheries Board of Canada* 28:1309–1318; for a human connection, see Stenhouse, M. J., and M. S. Baxter. 1977. "Bomb ^{14}C as a biological tracer." *Nature* 267:828–832.

5. Cailliet and Andrews, "Age-validated longevity of fishes," p. 105.

6. http://en.wikipedia.org/wiki/Yelloweye_rockfish.

7. http://www.conservationmagazine.org/2008/09/impostor-fish/.

8. Cailliet and Andrews, "Age-validated longevity of fishes," figure 3.

9. Andrews, A. H., G. M. Caillet, K. H. Coale, K. M. Munk, M. M. Mahoney, and V. M. O'Connell. 2002. "Radiometric age validation of the yelloweye rockfish (*Sebastes ruberrimus*) from southeastern Alaska." *Marine and Freshwater Research* 53:139–146.

10. Palumbi, S. R. 2004. "Fisheries science: Why mothers matter." *Nature* 430:621–622. http://palumbi.stanford.edu/manuscripts/Palumbi%202004a.pdf.

11. http://en.wikipedia.org/wiki/Yelloweye_rockfish.

12. Burton, E. J., A. H. Andrews, K. H. Coale, and G. M. Cailliet. 1999. "Application of radiometric age determination to three long-lived fishes using ^{210}Pb:^{226}Ra disequilibria in calcified structures: A review." In J. A. Musick (ed.). *Life in the Slow Lane: Ecology and Conservation of Long-Lived Marine Animals*. Special Publication 23. Bethesda, MD: American Fisheries Society, pp. 77–87.

13. http://www.youtube.com/watch?v=6EVajpR95bI, timestamp 0:55–1:25.

14. http://www.youtube.com/watch?v=6EVajpR95bI, timestamp 2:30–3:30.

15. http://seagrant.uaf.edu/news/96ASJ/05.06.96_BowheadAge.html; see also Noongwook, G., H. P. Huntington, and J. C. George. 2007. "Traditional knowledge of the bowhead whale (*Balaena mysticetus*) around St. Lawrence Island, Alaska." *Arctic* 60:47–54.

16. http://iwcoffice.org/conservation/lives.htm.

17. http://animals.nationalgeographic.com/animals/mammals/right-whale/.

18. John, C., and J. R. Bockstoce. 2008. "Two historical weapon fragments as an aid to estimating the longevity and movements of bowhead whales." *Polar Biology* 31:751–754.

19. http://news.nationalgeographic.com/news/2006/07/060713-whale-eyes_2.html.

20. George, J. C., J. Bada, J. Zeh, L. Scott, S. E. Brown, et al. 1999. "Age and growth estimates of bowhead whales (*Balaena mysticetus*) via aspartic acid racemization." *Canadian Journal of Zoology* 77:571–580; Rosa, C., J. Zeh, G. J. Craig, O. Botta, M. Zauscher, et al. 2012. "Age estimates based on aspartic acid racemization for bowhead whales (*Balaena mysticetus*) harvested in 1998–2000 and the relationship between racemization rate and body temperature." *Marine Mammal Science* 29:424–445.

21. http://iwcoffice.org/conservation/status.htm.

22. Clapham, P., S. Young, and R. Brownell Jr. 1999. *Baleen Whales: Conservation Issues and the Status of the Most Endangered Populations.* Paper 104. Washington, DC: U.S. Department of Commerce. http://digitalcommons.unl.edu/usdeptcommercepub/104.

23. Lubetkin, S. C., J. E. Zeh, C. Rosa, and J. C. George. 2008. "Age estimation for young bowhead whales (*Balaena mysticetus*)." *Canadian Journal of Zoology* 86:525–538. doi: 10.1139/Z08-028.

24. http://www.demogr.mpg.de/longevityrecords/0303.htm.

25. Medawar, P. 1957. *An Unsolved Problem in Biology.* London: H. K. Lewis and Co.

26. Zug, G. R., and J. F. Parham. 1996. "Age and growth in leatherback turtles, *Dermochelys coriacea* (Testudines: Dermochelyidae): A skeletochronological analysis." *Chelonian Conservation Biology* 2:244–249.

27. Eckert, K. L., and C. Luginbuhl. 1988. "Death of a giant." *Marine Turtle Newsletter* 43:2–3.

28. Shine, R., and J. B. Iverson. 1995. "Patterns of survival, growth and maturation in turtles." *Oikos* 72:343–348.

29. Zug, G. R., G. H. Balazs, J. A. Wetherall, D. M. Parker, K. Shawn, and K. Murakavua. 2002. "Age and growth of Hawaiian green seaturtles (*Chelonia mydas*): An analysis based on skeletochronology." *Fishery Bulletin* 100:117–127.

30. Elgar, M. A., and L. J. Heaphy. 1989. "Covariation between clutch size, egg weight and egg shape: Comparative evidence for chelonians." *Journal of Zoology* 219:137–152.

31. For leatherback sea turtles, one study estimated that the average female returns to her natal beach only every 3.2 years; Reina, R. D., P. A. Mayor, J. R. Spotila, R. Piedra, and F. V. Paladino. 2002. "Nesting ecology of the leatherback turtle, *Dermochelys coriacea,* at Parque Nacional Marino Las Baulas, Costa Rica: 1988–1989 to 1999–2000." *Copeia* 2002:653–664.

32. Chaloupka, M., and C. Limpus. 2002. "Survival probability estimates for the endangered loggerhead sea turtle resident in southern Great Barrier Reef waters." *Marine Biology* 140:267–277.

33. Carr, A. 1987. "New perspectives on the pelagic stage of sea turtle development." *Conservation Biology* 1:103–121.

34. Reported by N. Angier in 2006 at http://www.nytimes.com/2006/12/12/science/12turt.html.

35. http://oceanexplorer.noaa.gov/explorations/06laserline/background/blackcoral/blackcoral.html.

36. http://www.sciencecodex.com/stanford_researchers_say_living_corals_thousands_of_years_old_hold_clues_to_past_climate_changes_0.

37. http://blogs.sciencemag.org/newsblog/2008/02/methuselah-of-t.html.

38. Roark, E. B., T. P. Guilderson, R. B. Dunbar, S. J. Fallon, and D. A. Mucciarone. 2009. "Extreme longevity in proteinaceous deep-sea corals." *Proceedings of the National Academy of Sciences, USA* 106:5204–5208. doi: 10.1073/pnas.0810875106.

39. http://news.discovery.com/earth/caribbean-black-coral-date-back-to-jesus-110405.html.

40. Piraino, S., D. De Vito, J. Schmich, J. Bouillon, and F. Boero. 2004. "Reverse development in Cnidaria." *Canadian Journal of Zoology* 82:1748–1754.

41. Piraino, S., F. Boero, B. Aeschbach, and V. Scmid. 1996. "Reversing the life cycle: Medusae transforming into polyps and cell transdifferentiation in *Turritopsis nutricula* (Cnidaria, Hydrozoa)." *Biology Bulletin* 190:302–312.

42. http://blogs.discovermagazine.com/discoblog/2009/01/29/the-curious-case -of-the-immortal-jellyfish/.

43. http://news.nationalgeographic.com/news/2009/01/090130-immortal -jellyfish-swarm.html.

Chapter Seven: The Fastest Sprints and Longest Journeys

1. See the list of fish swimming speeds at FishBase: http://www.fishbase.org/ Topic/List.php?group=32.

2. Ellis, R. 2013. *Swordfish: A Biography of the Ocean Gladiator.* Chicago: University of Chicago Press.

3. Lee, H. J., Y. J. Jong, L. M. Chang, and W. L. Wu. 2009. "Propulsion strategy analysis of high-speed swordfish." *Transactions of the Japan Society for Aeronautical and Space Sciences* 52:11–20.

4. De Sylva, D. P. 1957. "Studies on the age and growth of the Atlantic sailfish, *Istio-phorus americanus* (Cuvier), using length-frequency curves." *Bulletin of Marine Science* 7:1–20.

5. http://en.wikipedia.org/wiki/Sailfish.

6. Block, B., D. Booth, and F. G. Carey. 1992. "Direct measurement of swimming speeds and depth of blue marlin." *Journal of Experimental Biology* 166:267–284.

7. And memory! Their records were all lost in a fire, but the Long Key fishermen remembered that it took 3 seconds for 100 yards of line to reel out. See Ellis, *Sword-fish*, p. 156.

8. Walters, V., and H. L. Fierstine. 1964. "Measurements of swimming speeds of yellowfin tuna and wahoo." *Nature* 202:208–209.

9. http://seagrant.gso.uri.edu/factsheets/swordfish.html.

10. Carcy, F. G., J. M. Teal, J. W. Kanwisher, K. D. Lawson, and J. S. Beckett. 1971. "Warm-bodied fish." *American Zoologist* 11:137–143.

11. Agris, P. F., and I. D. Campbell. 1979. "A brain heater in the swordfish." *Science* 205:160; Block, B. A. 1987. "Billfish brain and eye heater: A new look at non-shivering heat production." *Physiology* 2:208–213; Block, B. A. 1986. "Structure of the brain and eye heater tissue in marlins, sailfish, and spearfishes." *Journal of Morphology* 190:169–189; see also http://greenrage.wordpress.com/2008/05/16/ anatomy-week-brain-heaters-in-marlins-and-sailfish/.

12. Fritsches, K. A., R. W. Brill, and E. J. Warrant. 2005. "Warm eyes provide superior vision in swordfishes." *Current Biology* 15:55–58.

13. http://www.fishbase.org.

14. Denny, M. W. 1993. *Air and Water: The Biology and Physics of Life's Media.* Princeton, NJ: Princeton University Press.

15. http://www.discoverlife.org/20/q?search=Exocoetus+volitans&b=FB1032.

16. Davenport, J. 1992. "Wing loading, stability, and morphometric relationships in flying fish (Exocoetidae) from the North Eastern Atlantic." *Journal of the Marine Biology Association, UK* 72:25–39.

17. http://www.montereybayaquarium.org/animals/AnimalDetails. aspx?enc=VsGX+Lst7QZT1ija0iwiEA.

18. Fish, F. E. 1990. "Wing design and scaling of flying fish with regard to flight performance." *Journal of Zoology, London* 221:391–403.

19. Oxenford, H. A., and W. Hunte. 1999. "Feeding habits of the dolphinfish (*Coryphaena hippurus*) in the eastern Caribbean." *Scientia Marina* 63:303–315. doi: 10.3989/scimar.1999.63n3-4317.

20. Au, D., and D. Weihs. 1980. "At high speeds dolphins save energy by leaping." *Nature* 284:548–550.

21. http://en.wikipedia.org/wiki/Humpback_whale.

22. http://en.wikipedia.org/wiki/Fin_whale.

23. Clapham, P. J., and J. G. Mead. 1999. "*Megaptera novaeangliae*." *Mammalian Species* 604:1–9.

24. Fish, F. E., L. E. Howle, and M. M. Murray. 2008. "Hydrodynamic flow control in marine mammals." *Integrative and Comparative Biology* 48:788–800.

25. http://www.nextenergynews.com/news1/next-energy-news3.7b.html.

26. http://www.gizmag.com/bumpy-whale-fins-set-to-spark-a-revolution-in -aerodynamics/9020/.

27. Castro, P., and M. Huber. 2000. *Marine Biology*, third edition. Boston: McGraw-Hill, pp. 119–121.

28. http://lyle.smu.edu/~pkrueger/propulsion.htm.

29. http://en.wikipedia.org/wiki/Squid_giant_axon.

30. O'Dor, R., J. Stewart, W. Gilly, J. Payne, T Cerveira Borges, and T. Thys. 2012. "Squid rocket science: How squid launch into air." *Deep Sea Research Part II: Topical Studies in Oceanography.* http://dx.doi.org/10.1016/j.dsr2.2012.07.002.

31. Muramatsu, K., J. Yamamoto, T. Abe, K. Sekiguchi, N. Hoshi, and Y. Sakurai. 2013. "Oceanic squid do fly." *Marine Biology* 160:1171–1175.

32. http://www.gma.org/lobsters/allaboutlobsters/society.html; http://slgo.ca/ en/lobster/context/foodchain.html.

33. http://marinebio.org/species.asp?id=533.

34. Nauen, J. C., and R. E. Shadwick. 1999. "The scaling of acceleratory aquatic locomotion: Body size and tail-flip performance of the California spiny lobster *Panulirus interruptus*." *Journal of Experimental Biology* 202:3181–3193. http://jeb.biologists .org/cgi/reprint/202/22/3181.pdf.

35. Of course, if your Bugati can only accelerate for a second, like a lobster does, then you should consider returning it to the dealer; http://en.wikipedia.org/wiki/ List_of_fastest_production_cars_by_acceleration.

36. Edwards, D. H., W. J. Heitler, and F. B. Krasne. 1999. "Fifty years of a command neuron: The neurobiology of escape behavior in the crayfish." *Trends in Neurosciences* 22:153–161; or see http://en.wikipedia.org/wiki/Caridoid_escape_reaction.

37. Edwards et al., "Fifty years of a command neuron."

38. http://en.wikipedia.org/wiki/Caridoid_escape_reaction.

39. Wine, J. J., and F. B. Krasne. 1972. "The organization of escape behaviour in the crayfish." *Journal of Experimental Biology* 56:1–18.

40. http://en.wikipedia.org/wiki/Command_neuron.

41. http://en.wikipedia.org/wiki/Alpheidae.

42. Johnson, M. W., F. A. Everest, and R. W. Young. 1947. "The role of snapping shrimp (*Crangon* and *Synalpheus*) in the production of underwater noise in the sea." *Biological Bulletin* 93:122–138.

43. Versluis, M., B. Schmitz, A. von der Heydt, and D. Lohse. 2000. "How snapping shrimp snap: Through cavitating bubbles." *Science* 289:2114–2117; http://www.youtube.com/watch?v=XC6I8iPiHT8.

44. Lohse, D., B. Schmitz, and M. Versluis. 2001. "Snapping shrimp make flashing bubbles." *Nature* 413:477–478; see also http://news.nationalgeographic.com/news/2001/10/1003_SnappingShrimp.html.

45. Lohse et al., "Snapping shrimp make flashing bubbles."

46. See work by Emmett Duffy, such as Duffy, J. E. 2003. "The ecology and evolution of eusociality in sponge-dwelling shrimp." In T. Kikuchi, S. Higashi, and N. Azuma (eds.). *Genes, Behaviors, and Evolution in Social Insects.* Sapporo, Japan: University of Hokkaido Press, pp. 217–252. Emmett suggests you also take a look at this snippet from Blue Planet: http://www.youtube.com/watch?v=z735I4m8F8c.

47. Clapham, P. J. 2000. "The humpback whale." In J. Mann (ed.). *Cetacean Societies: Field Studies of Dolphins and Whales.* Chicago: University of Chicago Press, pp. 173–198.

48. Rice, D. W., A. A. Wolman, and H. W. Braham. 1984. "The gray whale, *Eschrichtius robustus.*" *Marine Fisheries Review* 46(4):7–14.

49. http://www.npr.org/blogs/thesalt/2012/07/24/157317262/how-many-calories-do-olympic-athletes-need-it-depends.

50. For a study of blue whale migrations, see Mate, B. R., B. A. Lagerquist, and J. Calambokidis. 1999. "Movements of North Pacific blue whales during the feeding season off southern California and their southern fall migration." *Marine Mammal Science* 15:1246–1257.

51. Fish et al., "Hydrodynamic flow control"; http://icb.oxfordjournals.org/content/48/6/788.full.

52. Fish et al., "Hydrodynamic flow control."

53. See the blue whale facts at http://acsonline.org/fact-sheets/blue-whale-2/.

54. Grebmeier, J. M. 2012. "Shifting patterns of life in the Pacific Arctic and sub-Arctic Seas." *Marine Science* 4: 63–78.

55. Alter, S. E., E. Rynes, and S. R. Palumbi. 2007. "DNA evidence for historic population size and past ecosystem impacts of gray whales." *Proceedings of the National Academy of Sciences, USA* 104:15162–15167.

56. http://www.nature.com/news/2011/110504/full/473016a.html.

57. http://en.wikipedia.org/wiki/Albatross; see also http://youtu.be/MBAr_aGaGA8?t=1m26s.

58. See "Grey-headed albatross," http://youtu.be/sUJx_At0sug.

59. http://en.wikipedia.org/wiki/Wandering_Albatross.

60. Safina, C. 2002. *Eye of the Albatross: Visions of Hope and Survival.* New York: Henry Holt and Company.

61. Pennycuick, C. J. 1982. "The flight of petrels and albatrosses (Procellariiformes), observed in South Georgia and its vicinity." *Philosophical Transactions of the Royal Society of London B* 300:75–106.

62. Weimerskirch, H., T. Guionnet, J. Martin, S. A. Shaffer, and D. P. Costa. 2000. "Fast and fuel efficient? Optimal use of wind by flying albatrosses." *Proceedings of the Royal Society of London B* 267:1869–1874.

63. Rayleigh, J.W.S. 1883. "The soaring of birds." *Nature* 27:534–535; http://en .wikipedia.org/wiki/Dynamic_soaring.

64. Richardson, P. L. 2011. "How do albatrosses fly around the world without flapping their wings?" *Progress in Oceanography* 88:46–58.

65. Richardson, "How do albatrosses fly?" p. 56.

66. Lecomte, V. J., G. Sorci, S. Cornet, A. Jaeger, B. Faivre, et al. 2010. "Patterns of aging in the long-lived wandering albatross." *Proceedings of the National Academy of Sciences, USA* 107:6370–6375.

67. Coleridge, S. T. 1798. *The Rime of the Ancient Mariner,* Part II. Online at http://www.online-literature.com/coleridge/646.

Chapter Eight: The Hottest

1. Somero, G. N., and A. L. DeVries. 1967. "Temperature tolerance of some Antarctic fishes." *Science* 156:257–258.

2. We give Fahrenheit values throughout, though most scientific sources note temperature in degrees Celsius.

3. Lutz, R. A. 2012. "Deep-sea hydrothermal vents." In E. Bell (ed.). *Life at Extremes: Environments, Organisms and Strategies for Survival.* Wallingford, UK: CABI, pp. 242–270.

4. Desbruyères, D., and L. Laubier. 1980. "*Alvinella pompejana* gen. sp. nov., Ampharetidae aberrant des sources hydrothermales de la ride Est-Pacifique." *Oceanologica Acta* 3:267–274; Ravaux, J., G. Hamel, M. Zbinden, A. A. Tasiemski, I. Boutet, et al. 2013. "Thermal limit for metazoan life in question: In vivo heat tolerance of the Pompeii worm." *PLoS One* 8: e64074.

5. http://www.exploratorium.edu/aaas-2001/dispatches/thermal_worm.html.

6. Desbruyères and Laubier, "*Alvinella pompejana* gen. sp. nov."

7. Ravaux et al., "Thermal limit for metazoan life."

8. For a good open-source brief review, see Ravaux et al., "Thermal limit for metazoan life."

9. Jollivet, D., J. Mary, N. Gagnière, A. Tanguy, E. Fontanillas, et al. 2012. "Proteome adaptation to high temperatures in the ectothermic hydrothermal vent Pompeii worm." *PLoS One* 7:e31150. doi: 10.1371/journal.pone.0031150.

10. Of course, it has bacterial symbionts like the deep-sea tube worms, the shrimp discussed in the next section, and most animal life of the vents; for the sequencing plan, see http://www.jgi.doe.gov/sequencing/why/3135.html.

11. Ravaux et al. seem to have managed this feat.

12. http://www.untamedscience.com/biology/world-biomes/deep-sea-biome.

13. Next time you are near a comfortable campfire fire or fireplace, take a minute to feel the warmth on your face. Place an empty glass in front of your face—the warmth drops but is still there. Now fill the glass with water—the warmth is gone, absorbed by the water in the glass.

14. Castro, P., and M. Huber. 2000. *Marine Biology,* third edition. Boston: McGraw-Hill, p. 351.

15. Hügler, M., J. M. Petersen, N. Dubilier, J. F. Imhoff, and S. M. Sievert. 2011. "Pathways of carbon and energy metabolism of the epibiotic community associated with the deep-sea hydrothermal vent shrimp *Rimicaris exoculata*." *PLoS One* 6:e16018.

16. Van Dover, C. L., E. Z. Szuts, S. C. Chamberlain, and J. R. Cann. 1989. "A novel eye in 'eyeless' shrimp from hydrothermal vents of the Mid-Atlantic Ridge." *Nature* 337:458–460; http://deepseanews.com/2010/04/the-eye-of-the-vent-shrimp/.

17. O'Neill, P. J., R. N. Jinks, E. D. Herzog, B. A. Battelle, L. Kass, G. H. Renninger, and S. C. Chamberlain. 1995. "The morphology of the dorsal eye of the hydrothermal vent shrimp, *Rimicaris exoculata*." *Visual Neuroscience* 12:861–875.

18. Pelli, D. G., and S. C. Chamberlain. 1989. "The visibility of 350° C black-body radiation by the shrimp *Rimicaris exoculata* and man." *Nature* 337:460–461. http://www.nature.com/nature/journal/v337/n6206/abs/337460a0.html.

19. Reviewers and friends have sparked a philosophical discussion about this point far larger than the shrimp. Are these shrimp really blind? Sort of, though they see light. Do they have eyes? Well, yes, but not on their heads. Do they see? Well, yes, but they do not see images. If you close your eyes and try to navigate by the feeling of heat you can get from the back of your neck, will you be said to be stumbling around blindly? Readers are encouraged to decide for themselves whether the "rift-shrimp without eyes" is misnamed.

20. http://www.stanford.edu/group/microdocs/whatisacoral.html.

21. Castro and Huber, *Marine Biology*, pp. 282–284.

22. Wilkinson C. 2000. *Status of Coral Reefs of the World: 2000.* Townsville, Australia: Global Coral Reef Monitoring Network and Australian Institute of Marine Science.

23. Wilkinson, *Status of Coral Reefs*, pp. 104–105.

24. Grigg, R. W. 1982. "Darwin Point: A threshold for atoll formation." *Coral Reefs* 1:29–34.

25. Castro and Huber, *Marine Biology*, pp. 282–283.

26. Jokiel, P., and S. Coles. 1990. "Response of Hawaiian and other Indo-Pacific reef corals to elevated temperature." *Coral Reefs* 8.155–162.

27. Bleaching temperature is the mean monthly maximum temperature plus 1° C.

28. http://www.telegraph.co.uk/earth/earthnews/7896403/Coral-reefs-suffer -mass-bleaching.html

29. See the current map at http://www.osdpd.noaa.gov/ml/ocean/cb/dhw.html.

30. http://alumni.stanford.edu/get/page/magazine/article/?article_id=28770.

31. See a historical summary in Mergner, H. 1984. "The ecological research on coral reefs of the Red Sea." *Deep Sea Research A. Oceanographic Research Papers* 31:855–884.

32. Cantin, N. E., A. L. Cohen, K. B. Karnauskas, A. M. Tarrant, and D. C. McCorkle. 2010. "Ocean warming slows coral growth in the central Red Sea." *Science* 329:322–325. doi 10.1126/science.1190182. http://re.indiaenvironmentportal.org .in/files/Ocean%20warming.pdf; http://www.sciencedaily.com/releases/2010/07/ 100715152909.htm.

33. www.stanford.edu/group/microdocs/typesofreefs.htm.

34. Darwin, C. 1842. *The Structure and Distribution of Coral Reefs. Being the First Part of the Geology of the Voyage of the* Beagle, *under the Command of Capt. Fitzroy, R.N. during the Years 1832 to 1836.* London: Smith Elder and Co., chapter 6. http://www .readbookonline.net/read/63216/112102/.

35. http://www.britannica.com/EBchecked/topic/176462/East-African-Rift -System.

36. Roberts, C. M., A. R. Dawson Shepherd, and R.F.G. Ormond. 1992. "Large-scale variation in assemblage structure of Red Sea butterflyfishes and angelfishes." *Journal of Biogeography* 19:239–250.

37. http://www.richardfield.freeservers.com/newdir/butterfl.htm.

38. Hsu, K., J. Chen, and K. Shao. 2007. "Molecular phylogeny of *Chaetodon* (Teleostei: Chaetodontidae) in the Indo-West Pacific: Evolution in geminate species pairs and species groups." *Raffles Bulletin of Zoology Supplement* 14:77–78.

39. Lampert-Karako, S., N. Stambler, D. J. Katcoff, Y. Achituv, Z. Dubinsky, and N. Simon-Blecher. 2008. "Effects of depth and eutrophication on the zooxanthella clades of *Stylophora pistillata* from the Gulf of Eilat (Red Sea)." *Aquatic Conservation: Marine and Freshwater Ecosystems* 18:1039–1045. doi: 10.1002/aqc.927.

40. Cantin et al., "Ocean warming slows coral growth."

41. Cantin et al., "Ocean warming slows coral growth."

42. http://en.wikipedia.org/wiki/Gulf_of_California.

43. Brownell, R. L., Jr. 1986. "Distribution of the vaquita, *Phocoena sinus*, in Mexican waters." *Marine Mammal Science* 2:299–305.

44. http://vaquita.tv/documentary/introduction/.

45. http://swfsc.noaa.gov/textblock.aspx?Division=PRD&ParentMenuId=229&id =13812.

46. http://swfsc.noaa.gov/textblock.aspx?Division=PRD&ParentMenuId=229&id =13812.

47. Norris, K. S., and W. N. McFarland. 1958. "A new harbor porpoise of the genus *Phocoena* from the Gulf of California." *Journal of Mammalogy* 39:22–39; http://www .iucn-csg.org/index.php/vaquita/.

48. Rosel, P. E., M. G. Haygood, and W. F. Perrin. 1995. "Phylogenetic relationships among the true porpoises (Cetacea: Phocoenidae)." *Molecular Phylogenetics and Evolution* 4:463–474; http://en.wikipedia.org/wiki/Spectacled_porpoise.

49. Somero, G. N., and A. L. DeVries. 1967. "Temperature tolerance of some Antarctic fishes." *Science* 156:257–258.

50. Jollivet et al., "Proteome adaptation to high temperatures."

51. Stillman, J. H. 2003. "Acclimation capacity underlies susceptibility to climate change." *Science* 301:65.

Chapter Nine: The Coldest

1. Daston, L., and K. Park. 2001. *Wonders and the Order of Nature, 1150–1750*. New York: Zone Books, pp. 255–302.

2. http://www.narwhal.org/NarwhalIntro.html.

3. Heide-Jørgensen, M. P., and K. L. Laidre. 2006. *Greenland's Winter Whales: The Beluga, the Narwhal and the Bowhead Whale*. B. M. Jespersen (ed.). Nuuk, Greenland: Ilinniusiorfik Undervisningsmiddelforlag, pp. 100–125.

4. Laidre, K. L., and Heide-Jørgensen, M. P. 2005. "Winter feeding intensity of narwhals (*Monodon monoceros*)." *Marine Mammal Science* 21:45–57; http://www.britannica .com/blogs/2011/03/legend-mystery-narwhal/. Both references also summarize local narwhal lore.

5. Daston and Park, *Wonders and the Order of Nature.*

6. http://acsonline.org/fact-sheets/narwhal/.

7. Silverman, H. B., and M. J. Dunbar. 1980. "Aggressive tusk use by the narwhal (*Monodon monoceros*)." *Nature* 284:56–57.

8. The same could be said of sports cars, or course.

9. Laidre, K. L., M. P. Heide-Jørgensen, O. A. Jørgensen, and M. A. Treble. 2004. "Deep-ocean predation by a high Arctic cetacean. *ICES Journal of Marine Science* 61:430–440. doi: 10.1016/j.icesjms.2004.02.002, http://icesjms.oxfordjournals.org/content/61/3/430.full.

10. Laidre et al., "Deep-ocean predation."

11. Laidre, K. L., M. P. Heide-Jorgensen, R. Dietz, R. C. Hobbs, and O. A. Jørgensen. 2003. "Deep-diving by narwhals *Monodon monoceros*: Differences in foraging behavior between wintering areas?" *Marine Ecology Progress Series* 261:269–281. http://www.int-res.com/abstracts/meps/v261/p269-281/.

12. http://smithsonianscience.org/2012/03/new-fossil-whale-species-raises-mystery-regarding-why-narwhals-and-belugas-live-only-in-cold-water/.

13. http://smithsonianscience.org/2012/03/new-fossil-whale-species-raises-mystery-regarding-why-narwhals-and-belugas-live-only-in-cold-water/.

14. Wilson, D. E., M. A. Bogan, R. L. Brownell Jr., A. M. Burdin, and M. K. Maminov. 1991. "Geographic variation in sea otters, *Enhydra lutris*." *Journal of Mammalogy* 72:22–36.

15. http://en.wikipedia.org/wiki/Kelp.

16. Williams, T. D., D. D. Allen, J. M. Groff, and R. L. Glass. 1992. "An analysis of California sea otter (*Enhydra lutris*) pelage and integument." *Marine Mammal Science* 8:1–18.

17. Palumbi, S. R., and C. Sotka. 2010. *The Death and Life of Monterey Bay: A Story of Revival.* Washington, DC: Island Press, chapter 2.

18. Palumbi and Sotka, *The Death and Life of Monterey Bay,* chapter 2.

19. Palumbi and Sotka, *The Death and Life of Monterey Bay,* chapter 2.

20. Palumbi and Sotka, *The Death and Life of Monterey Bay,* chapter 2.

21. http://osprey.bcodmo.org/project.cfm?id=188&flag=view.

22. Bolin, R. L. 1938. "Reappearance of the southern sea otter along the California coast." *Journal of Mammalogy* 19:301–303.

23. Bolin, "Reappearance of the southern sea otter."

24. Palumbi and Sotka, *The Death and Life of Monterey Bay,* chapter 2.

25. Duggins, D. O. 1980. "Kelp beds and sea otters: An experimental approach." *Ecology* 61:447–453. http://dx.doi.org/10.2307/1937405.

26. http://en.wikipedia.org/wiki/Southern_Ocean; Castro, P., and M. Huber. 2000. *Marine Biology,* third edition. Boston: McGraw-Hill.

27. Logsdon Jr., J. M., and W. F. Doolittle. 1997. "Origin of antifreeze protein genes: A cool tale in molecular evolution." *Proceedings of the National Academy of Sciences, USA* 94:3485–3487. http://www.pnas.org/content/94/8/3485.full.

28. http://www.msnbc.msn.com/id/13426864/ns/technology_and_science-science/.

29. Lodgson and Doolittle, "Origin of antifreeze protein genes."

30. Knight, C. A., A. L. De Vries, L. D. Oolman, L. D. 1984. "Fish antifreeze protein and the freezing and recrystallization of ice." *Nature* 308:295–296. http://www.nature.com/nature/journal/v308/n5956/abs/308295a0.html.

31. http://www.listener.co.nz/current-affairs/science/fishing-in-antarctica/.

32. Evans, C. W., V. Gubala, R. Nooney, D. E. Williams, M. A. Brimble, and A. L. Devries. 2011. "How do Antarctic notothenioid fishes cope with internal ice? A novel function for antifreeze glycoproteins." *Antarctic Science* 23:57–64.

33. http://www.exploratorium.edu/origins/antarctica/ideas/fish4.html.

34. Cheng, C.H.C. 1998. "Evolution of the diverse antifreeze proteins." *Current Opinion in Genetics and Development* 8:715–720.

35. http://www.rcsb.org/pdb/education_discussion/molecule_of_the_month/download/Antifreeze-Prot.pdf.

36. http://www.nytimes.com/2006/07/26/dining/26cream.html?_r=1.

37. http://www.food.gov.uk/multimedia/pdfs/ispfactsheet.

38. Macdonald, A., and C. Wunsch 1996. "An estimate of global ocean circulation and heat fluxes." *Nature* 382:436–439.

39. Smith, R. C., D. G. Martinson, S. E. Stammerjohn, R. A. Iannuzzi, and K. Ireson. 2008. "Bellingshausen and western Antarctic Peninsula region: Pigment biomass and sea-ice spatial/temporal distributions and interannual variabilty." *Deep Sea Research Part II: Topical Studies in Oceanography* 55:1949–1963.

40. Atkinson, A., V. Siegel, E. A. Pakhomov, M. J. Jessopp, and V. Loeb. 2009. "A re-appraisal of the total biomass and annual production of Antarctic krill." *Deep Sea Research Part I: Oceanographic Research Papers* 56:727–740.

41. http://en.wikipedia.org/wiki/Antarctic_krill.

42. http://www.coolantarctica.com/Antarctica%20fact%20file/wildlife/antarctic_animal_adaptations.htm.

43. Marschall, H. P. 1988. "The overwintering strategy of Antarctic krill under the pack ice of the Weddell Sea." *Polar Biology* 9:129–135.

44. http://www.afsc.noaa.gov/nmml/education/pinnipeds/crabeater.php; http://animaldiversity.ummz.umich.edu/site/accounts/information/Lobodon_carcinophaga.html; Klages, N., and V. Cockcroft. 1990. "Feeding behaviour of a captive crabeater seal." *Polar Biology* 10:403–404.

45. http://en.wikipedia.org/wiki/Trophic_level#Biomass_transfer_efficiency.

46. Alter, S. E., S. D. Newsome, and S. R. Palumbi. 2012. "Pre-whaling genetic diversity and population ecology in eastern Pacific grey whales: Insights from ancient DNA and stable isotopes." *PLoS One* 7:e35039.

47. Hilborn, R., T. A. Branch, B. Ernst, A. Magnusson, C. V. Minte-Vera, et al. 2003. "State of the world's fisheries." *Annual Review of Environment and Resources* 28:359–99; Clapham, P. J., and C. S. Baker. 2009. "Modern whaling." In W. F. Perrin B. Würsig, and J.G.M. Thewissen (eds.). *Encyclopedia of Marine Mammals,* second edition, volume 2. New York: Academic Press, pp. 1328–1332.

48. Fraser, W. R., W. Z. Trivelpiece, D. G. Ainley, and S. G. Trivelpiece. 1992. "Increases in Antarctic penguin populations: Reduced competition with whales or a loss of sea ice due to environmental warming?" *Polar Biology* 11:525–531.

49. http://www.mofa.go.jp/policy/economy/fishery/whales/iwc/minke.html.

50. C. Scott Baker, personal communication, April 2013.

51. Ruegg, K., E. Anderson, C. S. Baker, M. Vant, J. Jackson, and S. R. Palumbi. 2010. "Are Antarctic minke whales unusually abundant because of 20th century whaling?" *Molecular Ecology* 19:281–291.

52. http://www.lenfestocean.org/press-release/new-study-suggests-minke-whales -are-not-preventing-recovery-larger-whales-0.

53. Fraser et al., "Increases in Antarctic penguin populations."

54. "Power from the sea." *Popular Mechanics,* December 1930.

55. http://www.isla.hawaii.edu/komnet/studies.php.

56. http://www.energysavers.gov/renewable_energy/ocean/index.cfm/mytopic= 50010.

57. Othmer, D. F., and O. A. Roels. 1973. "Power, fresh water, and food from cold, deep sea water." *Science* 182:121–125. doi: 10.1126/science.182.4108.121.

58. War, J. C. 2011. "Land-based temperate species mariculture in warm tropical Hawaii." Oceans 2011 Conference Proceedings, Kona, Hawaii, September 19–22. http://ieeexplore.ieee.org/xpls/abs_all.jsp?arnumber=6107220&tag=1.

59. War, "Land-based temperate species mariculture."

60. http://www.energysavers.gov/renewable_energy/ocean/index.cfm/mytopic= 50010.

61. Barbeitos, M. S., S. L. Romano, and H. R. Lasker. 2010. "Repeated loss of coloniality and symbiosis in scleractinian corals." *Proceedings of the National Academy of Sciences, USA* 107:11877–11882.

62. Cartwright, P., and A. Collins. 2007. "Fossils and phylogenies: Integrating multiple lines of evidence to investigate the origin of early major metazoan lineages." *Integrative and Comparative Biology* 47:744–751.

63. http://www.mareco.org/khoyatan/spongegardens/introduction.

64. http://wsg.washington.edu/communications/seastar/stories/a_07.html.

65. Conway, K. W., M. Krautter, J. V. Barrie, and M. Neuweiler. 2001. "Hexactinellid sponge reefs on the Canadian continental shelf: A unique 'living fossil.'" *Geoscience Canada* 28(2):71–78.

66. Brümmer, F., M. Pfannkuchen, A. Baltz, T. Hauser, and V. Thiel. 2008. "Light inside sponges." *Journal of Experimental Marine Biology and Ecology* 367:61–64.

67. http://arctic.synergiesprairies.ca/arctic/index.php/arctic/article/viewFile/ 1636/1615.

68. http://en.wikipedia.org/wiki/Roald_Amundsen.

69. http://www.norway.org/aboutnorway/history/expolorers/amundsen/.

70. Vermeij, G. J. 1991. "Anatomy of an invasion: The Trans-Arctic Interchange." *Paleobiology* 17:281–307.

71. http://www.ncdc.noaa.gov/paleo/abrupt/data2.html.

72. http://en.wikipedia.org/wiki/Eemian; also called the Sangamonian interglacial stage in North America.

73. Bryant, P. J. 1995. "Dating remains of grey whales from the eastern North Atlantic." *Journal of Mammalogy* 76:857–861. doi: 10.2307/1382754. JSTOR 1382754.

74. http://www.earthtimes.org/nature/grey-whale-eastern-pacific/1978/.

75. http://www.nasa.gov/topics/earth/features/icesat-20090707r.html.

76. http://www.journalgazette.net/article/20120819/NEWS04/308199949.

Chapter Ten: The Strangest Family Lives

1. Fautin, D., and G. Allen. 1997. *Field Guide to Anemone Fishes and Their Host Sea Anemones,* second edition. Perth, Australia: Western Australian Museum.

2. Fricke, H., and S. Fricke. 1977. "Monogamy and sex change by aggressive dominance in coral reef fish." *Nature* 266:830–832.

3. Pietsch, T. W. 2009. *Oceanic Anglerfishes: Extraordinary Diversity in the Deep Sea.* Berkeley: University of California Press.

4. Saunders, B. 2012. *Discovery of Australia's Fishes: A History of Australian Ichthyology to 1930.* Collingwood, Australia: CSIRO Publishing.

5. Regan, C. T. 1925. "Dwarfed males parasitic on the females in oceanic anglerfishes (*Pediculati ceratioidea*)." *Proceedings of the Royal Society of London B* 97:386–400. doi: 10.1098/rspb.1925.0006. http://www.jstor.org/pss/1443462.

6. Pietsch, *Oceanic Anglerfishes.*

7. http://www.nature.com/nature/journal/v256/n5512/abs/256038a0.html.

8. Pietsch, *Oceanic Anglerfishes.*

9. Caspers, H. 1984. "Spawning periodicity and habitat of the palolo worm *Eunice viridis* (Polychaeta: Eunicidae) in the Samoan Islands." *Marine Biology* 79:229–236. doi: 10.1007/BF00393254. Note that the genus name *Palola* is more often used now— see http://invertebrates.si.edu/palola/science.html.

10. A voluble and charming narrative about the July swarm of palolo worms in Florida is in Mayer, A. G. 1909. "The annual swarming of the Atlantic palolo." In *Proceedings of the 7th International Congress of Zoology.* Stanford, CA: Carnegie Institution for Science, pp. 147–151.

11. Caspers, "Spawning periodicity and habitat."

12. Hofmann, D. K. 1974. "Maturation, epitoky and regeneration in the polychaete *Eunice siciliensis* under field and laboratory conditions." *Marine Biology* 25:149–161.

13. Stölting, K. N., and A. B. Wilson. 2007. "Male pregnancy in seahorses and pipefish: Beyond the mammalian model." *BioEssays* 29:884–896.

14. Casey, S. P., H. J. Hall, H. F. Stanley, and A. C. Vincent. 2004. "The origin and evolution of seahorses (genus *Hippocampus*): A phylogenetic study using the *cytochrome b* gene of mitochondrial DNA. *Molecular Phylogenetics and Evolution* 30:261–272.

15. For an engaging review of things seahorse, see Foster, S. J., and A.C.J. Vincent. 2004. "Life history and ecology of seahorses: Implications for conservation and management." *Journal of Fish Biology* 65:1–61.

16. The high-speed suction seahorses use to engulf prey is described in Bergert, B. A., and P. C. Wainwright. 1997. "Morphology and kinematics of prey capture in the syngnathid fishes *Hippocampus erectus* and *Syngnathus floridae*." *Marine Biology* 127:563–570.

17. Vincent, A. 1994. "The improbable seahorse." *National Geographic,* August.

18. http://news.nationalgeographic.com/news/2002/06/0614_seahorse_recov.html.

19. http://www.youtube.com/watch?v=e8EfAODDoRo.

20. http://www.independent.co.uk/environment/sex-life-of-a-seahorse-413329 .html.

21. http://www.sciencenews.org/pages/pdfs/data/2000/157-11/15711-09.pdf.

22. Watch a seahorse giving birth at http://www.youtube.com/watch?v= uKrkXXaRMUI&NR.

23. Kvarnemo, C., G. I. Moore, A. G. Jones, W. S. Nelson, and J. C. Avise. 2000. "Monogamous pair bonds and mate switching in the Western Australian seahorse *Hippocampus subelongatus*." *Journal of Evolutionary Biology* 13:882–888.

24. Jones, A. G., G. I. Moore, C. Kvarnemo, D. Walker, and J. C. Avise. 2003. "Sympatric speciation as a consequence of male pregnancy in seahorses." *Proceedings of the National Academy of Sciences, USA* 100:6598–6603; Mattle, B., and A. B. Wilson. 2009. "Body size preferences in the pot-bellied seahorse *Hippocampus abdominalis*: Choosy males and indiscriminate females." *Behavioral Ecology and Sociobiology* 63:1403–1410.

25. http://academic.reed.edu/biology/courses/BIO342/2010_syllabus/2010_readings/berglund_2010.pdf; http://www.nature.com/news/2010/100317/full/news.2010.127.html; the original paper is Paczolt, K. A., and A. G. Jones. 2010. "Post-copulatory sexual selection and sexual conflict in the evolution of male pregnancy." *Nature* 464:401–404.

26. http://www.flmnh.ufl.edu/fish/Gallery/Descript/sergeantmajor/sergeantmajor.html; http://www.fishbase.us/summary/Abudefduf-vaigiensis.html.

27. http://animal.discovery.com/guides/fish/marine/damselintro.html.

28. A Richard Harris radio story from NPR is at http://www.npr.org/templates/story/story.php?storyId=111743524.

29. Foster, S. A. 1987. "Diel and lunar patterns of reproduction in the Caribbean and Pacific sergeant major damselfishes: *Abudefduf saxatilis* and *A. troschelii*." *Marine Biology* 95:333–343.

30. Gronell, A. M. 1989. "Visiting behaviour by females of the sexually dichromatic damselfish, *Chrysiptera cyanea* (Teleostei: Pomacentridae): A probable method of assessing male quality." *Ethology* 81:89–122.

31. Keenleyside, M. H. 1972. "The behaviour of *Abudefduf zonatus* (Pisces, pomacentridae) at Heron Island, Great Barrier Reef." *Animal Behaviour* 20:763–774.

32. Hoelzer, G. A. 1992. "The ecology and evolution of partial-clutch cannibalism by paternal Cortez damselfish." *Oikos* 65:113–120.

33. Castro, P., and M. Huber. 2000. *Marine Biology,* third edition. Boston: McGraw-Hill, pp. 173–175.

34. Robinson, P. W., D. P. Costa, D. E. Crocker, J. P. Gallo-Reynoso, C. D. Champagne, et al. 2012. "Foraging behavior and success of a mesopelagic predator in the northeast Pacific Ocean: Insights from a data rich species, the northern elephant seal." *PLoS One* 7:e36728.

35. http://www.parks.ca.gov/?page_id=1115.

36. Le Boeuf, B. J., R. Condit, P. A. Morris, and J. Reiter. 2011. "The northern elephant seal (*Mirounga angustirostris*) rookery at Año Nuevo: A case study in colonization." *Aquatic Mammals* 37:486–501. doi: 10.1578/AM.37.4.2011.486.

37. http://www.parks.ca.gov/?page_id=1115.

38. Le Boeuf, B. J., and R. S. Peterson. 1969. "Social status and mating activity in elephant seals." *Science* 163:91–93. doi: 10.1126/science.163.3862.91.

39. http://www.marinebio.net/marinescience/05nekton/esrepro.htm.

40. Microsoft Encarta Online Encyclopedia. 2009. "Elephant seal." http://encarta.msn.com.

41. Sanvito, S., F. Galimberti, and E. H. Miller. 2007. "Having a big nose: Structure, ontogeny, and function of the elephant seal proboscis." *Canadian Journal of Zoology* 85:207–220.

42. Le Boeuf, B. J. 1974. "Male-male competition and reproductive success in elephant seals." *American Naturalist* 14:163–176. http://mirounga.ucsc.edu/leboeuf/pdfs/malemalecompetition.1974.pdf.

43. "Henry IV, Part 2," Act 3, Scene 1, Hal's opening soliloquy.

44. Le Boeuf, "Male-male competition."

45. Le Boeuf, B. J., and J. Reiter. 1988. "Lifetime reproductive success in northern elephant seals." In T. H. Clutton-Brock (ed.). *Reproductive Success: Studies of Individual Variation in Contrasting Breeding Systems*. Chicago: University of Chicago Press, pp. 344–362.

46. Le Boeuf and Peterson, "Social status and mating activity in elephant seals."

47. http://www.royalbcmuseum.bc.ca/school_programs/octopus/index-part2.html.

48. http://bioweb.uwlax.edu/bio203/s2012/kalupa_juli/reproduction.htm.

49. Hochner B., T. Flash, C. Angisola, and L. Zullo. 2009. "Nonsomatotopic organization of the higher motor centers in octopus." *Current Biology* 19:1632–1636. http://www.ncbi.nlm.nih.gov/pubmed/19765993.

50. A motto from O'Dor, R. K., and D. M. Webber. 1986. "The constraints on cephalopods: Why squid aren't fish." *Canadian Journal of Zoology* 64:1591–1605.

51. Wodinsky, J. 1977. "Hormonal inhibition of feeding and death in octopus: Control by optic gland secretion." *Science* 198:948–951.

52. Anderson, R. C., J. B. Wood, and R. A. Byrne. 2002. "Octopus senescence: The beginning of the end." *Journal of Applied Animal Welfare Science* 5:275–283.

53. http://en.wikipedia.org/wiki/Tunicate;http://en.wikipedia.org/wiki/Botryllus_schlosseri.

54. Stoner, D. S., B. Rinkevich, and I. L. Weissman. 1999. "Heritable germ and somatic cell lineage competitions in chimeric colonial protochordates." *Proceedings of the National Academy of Sciences, USA* 96:9148–9153.

55. Oren, M., M.-L. Escande, G. Paz, Z. Fishelson, and B. Rinkevich. 2008. "Urochordate histoincompatible interactions activate vertebrate-like coagulation system components." *PLoS One* 3:e3123. doi: 10.1371/journal.pone.0003123.

56. Bancroft, F. W. 1903. "Variation and fusion of colonies in compound ascidians." *Proceedings of the California Academy of Sciences* 3:137–186.

57. Rinkevich, B., and I. L. Weissman. 1992. "Allogeneic resorption in colonial protochordates: Consequences of nonself recognition." *Developmental and Comparative Immunology* 16:275–286.

58. Oren et al., "Urochordate histoincompatible interactions"; see also Scofield, V. 1997. "Sea squirt immunity: The AIDS connection." *MBL Science,* winter 1988–1989. http://hermes.mbl.edu/publications/pub_archive/Botryllus/Botryllus.revised.html.

59. Carpenter, M. A., J. H. Powell, K. J. Ishizuka, K. J. Palmeri, S. Rendulic, and A. W. De Tomaso. 2011. "Growth and long-term somatic and germline chimerism following fusion of juvenile *Botryllus schlosseri*." *Biological Bulletin* 220:57–70; Stoner et al., "Heritable germ and somatic cell lineage competitions."

60. L. Tolstoy. *Anna Karenina*, p. 1.

61. Bobko, S. J., and S. A. Berkeley. 2004. "Maturity, ovarian cycle, fecundity, and age-specific parturition of black rockfish (*Sebastes melanops*)." *Fishery Bulletin* 102:418–429.

62. O'Dor and Webber, "The constraints on cephalopods."

63. Berkeley, S. A., C. Chapman, and S. M. Sogard. 2004. "Maternal age as a determinant of larval growth and survival in a marine fish, *Sebastes melanops*." *Ecology* 85:1258–1264.

64. Marshall, D. J., S. S. Heppell, S. B. Munch, and R. R. Warner. 2010. "The relationship between maternal phenotype and offspring quality: Do older mothers really produce the best offspring?" *Ecology* 91:2862–2873. http://dx.doi.org/10.1890/09-0156.1.

65. See the discussion in Palumbi, S. R. 2002. *The Evolution Explosion*. New York: W. W. Norton.

66. Some of the unexpected subleties of the sperm economy in fish are discussed in Warner, R. R. 1997. "Sperm allocation in coral reef fishes." *BioScience* 47:561–564.

67. See especially Ghiselin, M. T. 1974. *The Economy of Nature and the Evolution of Sex*. Berkeley: University of California Press.

Chapter Eleven: Future Extremes

1. http://www.genomenewsnetwork.org/articles/08_03/hottest.shtml.

2. Donner, S. D. 2009. "Coping with commitment: Projected thermal stress on coal reefs under different future scenarios." *PLoS One* 4:e5712. The new 2013 IPPC report lists the most recent climate predictions: http://www.climatechange2013.org/images/uploads/WGIAR5_WGI-12Doc2b_FinalDraft_Chapter11.pdf.

3. http://www.nature.com/news/ancient-migration-coming-to-america-1.10562.

4. http://opinionator.blogs.nytimes.com/2012/12/29/the-power-of-a-hot-body/; or just watch *The Matrix* again.

5. http://www.huffingtonpost.com/2011/11/16/calories-cold-weather_n_1096331.html.

6. Johnson, A. N., D. F. Cooper, and R.H.T. Edwards. 1977. "Exertion of stairclimbing in normal subjects and in patients with chronic obstructive bronchitis." *Thorax* 32:711–716. http://thorax.bmj.com/content/32/6/711.full.pdf.

7. Stillman, J. 2003. "Acclimation capacity underlies susceptibility to climate change." *Science* 301:65. doi: 10.1126/science.1083073.

8. Donner, S. D., W. J. Skirving, C. M. Little, M. Oppenheimer, and O. Hoegh-Guldberg. 2005. "Global assessment of coral bleaching and required rates of adaptation under climate change." *Global Change Biology* 11:2251–2265.

9. Cheung, W. W., V. W. Lam, J. L. Sarmiento, K. Kearney, R.E.G. Watson, et al. 2010. "Large-scale redistribution of maximum fisheries catch potential in the global ocean under climate change." *Global Change Biology* 16:24–35.

10. http://co2now.org/Current-CO2/CO2-Now/global-carbon-emissions.html.

11. http://www.epa.gov/climatechange/images/indicator_downloads/acidit-download1-2012.png.

12. pH is a measure of the concentration of hydrogen ions in water, a measure of acidity. The scale runs from 0 to 14, with 7 being neutral, and lower numbers indicating logarithmically increasing concentrations of acid. This means that a decrease of one unit in pH corresponds to ten times more acidity.

13. Orr, J. C., V. J. Fabry, O. Aumont, L. Bopp, S. C. Doney, et al. 2005. "Anthropogenic ocean acidification over the twenty-first century and its impact on calcifying organisms." *Nature* 437:681–686.

14. Palmer, A. R. 1992. "Calcification in marine molluscs: How costly is it?" *Proceedings of the National Academy of Sciences, USA* 89:1379–1382.

15. Cohen, A. L., and M. Holcomb. 2009. "Why corals care about ocean acidification: Uncovering the mechanism." *Oceanography* 22:118–127.

16. Kroeker, K., R. L. Kordas, R. N. Crim, and G. G. Singh. 2010. "Meta-analysis reveals negative yet variable effects of ocean acidification on marine organisms." *Ecology Letters* 13:1419–1434.

17. Lannig, G., S. Eilers, H. O. Pörtner, I. M. Sokolova, and C. Bock. 2010. "Impact of ocean acidification on energy metabolism of oyster, *Crassostrea gigas*—Changes in metabolic pathways and thermal response." *Marine Drugs* 8:2318–2339.

18. Kroeker et al., "Meta-analysis reveals negative yet variable effects."

19. Doney, S. C., V. J. Fabry, R. A. Feely, and J. A. Kleypas. 2009. "Ocean acidification: The other CO_2 problem." *Annual Review of Marine Science* 1:169–192. doi: 10.1146/annurev.marine.010908.163834.

20. http://www.nber.org/digest/nov06/w12159.html.

21. See http://www.sciencedaily.com/releases/2012/04/120411132219.htm; Barton, A., B. Hales, G. G. Waldbusser, C. Langdon, and R. A. Feely. 2012. "The Pacific oyster, *Crassostrea gigas*, shows negative correlation to naturally elevated carbon dioxide levels: Implications for near-term ocean acidification effects." *Limnology and Oceanography* 57:698–710; Hettinger, A., E. Sanford, T. M. Hill, A. D. Russell, K. N. Sato, et al. 2012. "Persistent carry-over effects of planktonic exposure to ocean acidification in the Olympia oyster." *Ecology* 93:2758–2768.

22. Edmunds, P. J. 2011. "Zooplanktivory ameliorates the effects of ocean acidification on the reef coral *Porites* spp." *Limnology and Oceanography* 56:2402.

23. Barshis, D. J., J. T. Ladner, T. A. Oliver, F. O. Seneca, N. Traylor-Knowles, and S. R. Palumbi. 2013. "Genomic basis for coral resilience to climate change." *Proceedings of the National Academy of Sciences, USA* 110:1387–1392.

24. Cazenave, A., and R. S. Nerem. 2004. "Present-day sea level change: Observations and causes." *Reviews of Geophysics* 42. doi: 10.1029/2003RG000139; Chen, J. L., C. R. Wilson, and B. D. Tapley. 2013. "Contribution of ice sheet and mountain glacier melt to recent sea level rise." *Nature Geoscience* 6:549–552.

25. Merrifield, M. A., S. T. Merrifield, and G. T. Mitchum. 2009. "An anomalous recent acceleration of global sea level rise." *Journal of Climate* 22:5772–5781; Vermeer, M., and S. Rahmstorf. 2009. "Global sea level linked to global temperature." *Proceedings of the National Academy of Sciences, USA* 106:21527–21532; see also Schaeffer, M., W. Hare, S. Rahmstorf, and M. Vermeer. 2012. "Long-term sea-level rise implied by 1.5° C and 2° C warming levels." *Nature Climate Change* 2:867–870.

26. Perrette, M., F. Landerer, R. Riva, K. Frieler, and M. Meinshausen. 2013. "A scaling approach to project regional sea level rise and its uncertainties." *Earth System Dynamics* 4:11–29.

27. Orson, R., W. Panageotou, and S. P. Leatherman. 1985. "Response of tidal salt marshes of the US Atlantic and Gulf coasts to rising sea levels." *Journal of Coastal Research* 1:29–37.

28. McGranahan, G., D. Balk, and B. Anderson. 2007. "The rising tide: Assessing the risks of climate change and human settlements in low elevation coastal zones." *Environment and Urbanization* 19:17–37.

29. Koch, E. W., E. B. Barbier, B. R. Silliman, D. J. Reed, G. M. Perillo, et al. 2009. "Non-linearity in ecosystem services: Temporal and spatial variability in coastal protection." *Frontiers in Ecology and the Environment* 7:29–37.

30. Danielsen, F., M. K. Sørensen, M. F. Olwig, V. Selvam, F. Parish, et al. 2005. "The Asian tsunami: A protective role for coastal vegetation." *Science* 310:643.

31. Sovacool, B. K. 2011. "Hard and soft paths for climate change adaptation." *Climate Policy* 11:1177–1183.

32. Roberts, C. 2007. *The Unnatural History of the Sea.* Washington, DC: Island Press.

33. Thurstan, R. H., and C. M. Roberts. 2010. "Ecological meltdown in the Firth of Clyde, Scotland: Two centuries of change in a coastal marine ecosystem." *PLoS One* 5:e11767. doi: 10.1371/journal.pone.0011767.

34. http://news.sciencemag.org/sciencenow/2010/05/british-trawlers-working-nearly-.html.

35. Pauly, D., V. Christensen, J. Dalsgaard, R. Froese, and F. Torres Jr. 1998. "Fishing down marine food webs." *Science* 279:860–863.

36. http://www.ehow.com/how_8255493_catch-scallops-holden-beach.html.

37. http://www.ncseagrant.org/home/coastwatch/coastwatch-articles?task=show Article&id=640.

38. http://en.wikipedia.org/wiki/Dead_zone_(ecology).

39. http://greenbizness.com/blog/wiki/chemical-fertilizer-use-in-usa/

40. http://www.noaanews.noaa.gov/stories2009/pdfs/new%20fact%20sheet%20dead%20zones_final.pdf.

41. http://www.guardian.co.uk/world/2009/aug/10/france-brittany-coast-seaweed-algae.

42. http://www.cdc.gov/nceh/hsb/hab/default.htm.

43. Diaz, R. J., and R. Rosenberg. 2008. "Spreading dead zones and consequences for marine ecosystems." *Science* 321:926–929. doi: 10.1126/science.1156401.

44. http://www.telegraph.co.uk/science/space/9125409/The-algae-bloom-so-big-it-can-be-seen-from-space.html.

45. http://phaeocystis.org/.

46. Fontaine, M. C., A. Snirc, A. Frantzis, E. Koutrakis, B. Öztürk, et al. 2012. "History of expansion and anthropogenic collapse in a top marine predator of the Black Sea estimated from genetic data." *Proceedings of the National Academy of Sciences, USA* 109:E2569–E2576.

47. Kideys, A. E. 2002. "Fall and rise of the Black Sea ecosystem." *Science* 297:1482–1484.

48. Daskalov, G. M., A. N. Grishin, S. Rodionov, and V. Mihneva. 2007. "Trophic cascades triggered by overfishing reveal possible mechanisms of ecosystem regime shifts." *Proceedings of the National Academy of Sciences, USA* 104:10518–10523.

49. http://en.wikipedia.org/wiki/Mnemiopsis_leidyi.

50. http://www.smithsonianmag.com/specialsections/40th-anniversary/Jellyfish-The-Next-Kings-of-the-Sea.html; Daskalov et al. "Trophic cascades triggered by overfishing."

51. Jackson, J.B.C. 1997. "Reefs since Columbus." *Coral Reefs* 16:23–32.

52. Jackson, J.B.C., M. X. Kirby, W. H. Berger, K. A. Bjorndal, L. W. Botsford, et al. 2001. "Historical overfishing and the recent collapse of coastal ecosystems." *Science* 293:629–637.

53. Pandolfi, J. M., J.B.C. Jackson, N. Baron, R. H. Bradbury, H. M. Guzman, et al. 2005. "Are US coral reefs on the slippery slope to slime?" *Science* 307:1725–1726.

54. Caldwell, M., A. Hemphill, T. C. Hoffmann, S. Palumbi, J. Teisch, and C. Tu. 2009. *Pacific Ocean Synthesis: Executive Summary.* Palo Alto, CA: Center for Ocean Solutions Publications, Stanford University. http://www.centerforoceansolutions.org/content/pacific-ocean-synthesis-executive-summary.

Epilogue: A Grand Bargain

1. Palumbi, S. R. 2001. "The ecology of marine protected areas." In M. D. Bertness, S. D. Gaines, and M. E. Hay (eds.). *Marine Community Ecology*. Sunderland, MA: Sinauer, pp. 509–530.

2. Alcala, A. C., G. R. Russ, A. P. Maypa, and H. P. Calumpong. 2005. "A long-term, spatially replicated experimental test of the effect of marine reserves on local fish yields." *Canadian Journal of Fisheries and Aquatic Science* 62:98–108.

3. This scenario is called RCP 8.5 by the Intergovernmental Panel on Climate Change. The IPCC's 2007 final report can be found online at http://www.aimes.ucar.edu/docs/IPCC.meetingreport.final.pdf. The 2013 report was not complete as of this writing, but the web site where it is being released is http://www.ipcc.ch.

Index

Page numbers for entries occurring in figures are followed by an *f* and those for entries in notes, by an *n*.

in coral reefs, 120
heat tolerance in, 112, 123
human impacts on (*See* overfishing)
lifespan of, 81–84, 88
in mangrove forests, 72–74
reproduction in, 141–49
saturated fat in, 56
speed of, 95–99
thermoregulation in, 96–97
fisheries. *See also* overfishing
lifespan of fish and, 82, 83–84
technological advances in, 166–67
"fishing down the food web," 167
flight
in albatross, 108–11, 110f
evolution of, 97
in flying fish, 97–99
flooding, sea level rise and, 163–64
flying fish, 97–99, 99f
food. *See* diet
food chains
fertilizers and, 169
future of, 173
krill in, 134–35
microbes in, 164, 169–70
overfishing and, 167–68, 170 72
fossil record. *See also specific species*
in Burgess Shale, 9–17
limitations of, 9, 12–13
fossils, living. *See* living fossils
France, microbial blooms in, 169
freezing point, of blood, 130–32
Friends of the Sea Otter, 130
frilled shark, 30–31, 31f, 33
fringing reefs, 119–20
Funafuti, sea level rise and, 163
fur, of sea otter, 128–29

Gallien, Brad, 66
game tape analogy, 15–17
gases, deep-sea pressure on, 54–55
genetics. *See also* DNA
of Cambrian Explosion, 12
of giant squid, 60
of heat tolerance, 118–19, 123
germ theory of disease, 37

Geukensia demissa (ribbed marsh mussel),
70
giant axon, 102
giant kelp, 128–29
giant Pacific octopus (*Enteroctopus dofleini*),
parental care in, 152
giant species, of deep sea, 57–61
giant squid (*Architeuthis*), 59–61
blood of, 3, 179n8
genetics of, 60
predation by, 60–61
as prey of sperm whale, 3–4, 60
size of, 3, 59
giant tube worms, 49–51, 58
gigantism, deep-sea, 57–61
gills, of horseshoe crabs, 25
glaciers, melting of, 163
glass sponges, cold tolerance of, 137–38
gliding, by albatross, 109–10, 110f
global warming. *See* climate change
glycoproteins, antifreeze, 131–32
goblin shark, 30, 30f, 34
Gould, Stephen Jay
on Burgess Shale species, 14, 15, 16
on chance, role of, 15–16
on diversity, peak of, 17
empty barrel theory of, 11–12, 17
game tape analogy of, 15–17
Wonderful Life, 15–16
gradients, intertidal, 67, 79–80
grasses, in salt marshes, 69–70
gray whale (*Eschrichtius robustus*)
migration by, 106, 108, 139–40
in Trans-Arctic Interchange, 139–40
Great Barrier Reef, 116
Great Britain, fisheries in, 166–67
Great Oxygenation Event, 6–8, 7f
Great Rift Valley, 119–20
green sea turtle (*Chelonia mydas*), lifespan
of, 87–88, 87f
Gulf of California, heat tolerance in, 121–23
Gulf of Mexico
agricultural run-off in, 169, 170
Deep Horizon oil platform explosion in,
39
Gynaecotyla, 69

lobstering, 103–4
lobsters, caridoid escape reaction in, 102,
    103–4
long-distance migrations, 106–11
    by albatross, 108–11, 110f
    by whales, 106–8, 139–40
Long Key Fishing Club, 96, 193n7
loosejaws, 63, 63f
luciferases, 62

MacFarland, William, 123
Magnetic North Pole, 139
Malacosteus niger (stoplight loosejaw), 63f
male pregnancy, in seahorses, 147–48
Malfatti, F., 185n26
mammals. *See also specific types*
    evolution of flight in, 97
mangrove forests, 70–74, 163
mantle
    of chambered nautilus, 22, 23
    of squid, 102
margarine, 56
marine protected areas, 176, 177
marine snow, 51
Marrella, 10
Maryland, horseshoe crabs in, 24
mass extinctions, 21, 31–32, 34, 45
maternal care. *See* parental care
mating behavior. *See* reproduction
Mediterranean Sea, whales in, 108
Megaptera novaeangliae (humpback whale)
    breaching by, 100–101
    fins of, 100–101, 101f
    lifespan of, 87–88
    migration by, 106
    size of, 100
Mesonychoteuthis hamiltoni (colossal squid),
    59–60
    as prey of sperm whale, 3–4
    size of, 59–60
metabolism
    in cold tolerance, 129
    microbial, 6–8, 37–40
    ocean acidity and, 162
    ocean temperature and, 160, 162
    oxygen in, 6–8, 11

Mexico, Gulf of
    agricultural run-off in, 169, 170
    Deep Horizon oil platform explosion in,
        39
microbes, 36–45. *See also specific types*
    in biomass of ocean, 37–38, 40, 43, 164
    blooms of (*See* microbial blooms)
    in chemistry of ocean, 36, 37–38
    definition of, 36
    discovery of, 36–37
    in empty barrel theory, 11
    in evolutionary history, 6–9, 11, 36
    future of, 172–73
    in germ theory of disease, 37
    human impacts on, 164–65, 168–70
    limitations on dominance of, 43–45
    metabolism of, 6–8, 37–40
    in microbial loop, 40–42, 43
    oxygen created by, 6–7
    population of, 36–38
    size of, 36–38
    types of, 36
    use of term, 185n26
microbial blooms
    agricultural run-off in, 169–70
    in Antarctica, 169–70
    chemistry of ocean during, 38
    damage caused by, 43, 169
    human impacts on, 168–70
    viruses and, 43–45
microbial loop, 40–42, 41f, 43
mid-intertidal zone, 74–76
migrations, 106–11
    by albatross, 108–11, 110f
    by whales, 106–8, 139–40
minke whales (*Balaenoptera*), hunting of,
    135
Mirounga angustirostris (northern elephant
    seal), reproduction in, 149–51
Mississippi River, agricultural run-off in,
    169
mitochondria, 8
mobile scavenger phase, 52
Molecular Clock, 12
mollusks. *See also specific types*
    in fossil record, 12

overfishing of, 170
species of, 123
Portuguese man-of-war, 78
potato bugs, 57
predation. *See also specific species*
coevolution in, 34
in intertidal zone, 68, 74–79
origins of, 12–13
speed and, 96
pregnancy, male, in seahorses, 147–48
pressure, water, in deep sea, 54–57
Princess Bride, The (film), 118
proboscis, 151
Prochlorococcus, 38–39, 43
productivity
in Antarctic, 135–36
of microbes, 164–65
Productivity Bomb, 165, 170–74, 177–78
Project Seahorse, 165–66
protected areas, marine, 176, 177
Protein Data Bank, 132
proteins
antifreeze, 131–32
viral, 42, 43
in whale eyes, 85
protists, in microbial loop, 41, 42
protozoa, size of, 36–37
Pyrolobus fumarii, 8–9

radulas, 68
rays, overfishing and, 167–68
RCP 2.6 scenario, 177f
RCP 8.5 scenario, 177f, 208n3
reactive oxygen species, 117
red (color), in deep sea, 63, 115
red mangroves, 71–72
Red Sea, corals in, 119–21
red snapper. *See* yellow-eye rockfish
red tides, 169, 170
reefs. *See also* coral reefs
cold-water, 137–38
Regan, Charles, 143
renewable energy, cold seawater as source
of, 136–37
reproduction, 141–57

in anglerfish, 142–44
in clownfish, 141–42
in damselfish, 148–49
in deep-sea gigantism, 59
diversity of strategies for, 155–57
in elephant seals, 149–51
lifespan and, 86, 88–90, 155–56
in octopuses, 152, 155
in palolo worm, 144–46
in seahorses, 147–48
in sea squirts, 152–55
Rescuers, The (film), 109
respiration
in mudskippers, 73–74
in sea urchins, 66–67
retinas, thermoregulation and, 97
Rhinoptera bonasus (cow-nosed ray), overfishing and, 167–68, 168f
Rhizophoraceae, 71–72
rhodopsin, 115
ribbed marsh mussel (*Geukensia demissa*),
70
Richardson, Phillip, 110–11
Riftia pachyptila, 49–51
rift shrimp (*Rimicaris exoculata*), perception
of light by, 114–15, 197n19
right handedness, 85
right whales, 84
Rimicaris exoculata (rift shrimp), perception
of light by, 114–15, 197n19
Roberts, Callum, 166–67
rockfish
lifespan of, 82–83, 155
reproduction in, 155, 156
Roels, Oswald, 137
roots
in mangrove forests, 71–72
in salt marshes, 69–70
Rouse, Greg, 54

sailfish (*Istiophorus*), speed of, 95–96,
193n7
salt
excretion of, 71–72
in mangrove forests, 71–72

wind, in albatross flight, 109–11, 110f
wind turbines, modeled on whale fins,
    100–101
wings
    of albatross, 109–10
    evolution of, 97
Wiwaxia, 10
wonder, need for sense of, vii
Wonderful Life (Gould), 15–16
worms. *See also specific types*
    in fossil record, 12–13

at hydrothermal vents, 49–51, 113–14,
    123

yellow-eye rockfish (*Sebastes ruberrimus*),
    83
yellowfin tuna, speed of, 96
yellow tides, 169

zombie bone worms, 53–54
zonation, intertidal, 67, 79–80
zooids, 152–55, 153f